To John with
— birthday. 77
— Ginny

WINDSINGER

WIND

SINGER

Gary M. Smith

Sierra Club Books

San Francisco

1976

The "power spot," on Orange Cliffs above the canyonlands, by Jim Conklin.

Hummingbird.

Frontispiece: Red Tail hawk, by Richard Howard.

Dick Smith tracking hawk above Chesler Park, in the canyonlands.

Charley Atkins on his Blue Ridge Mountain farm, Virginia.

Title Page: Fishhook Creek in the Sawtooths of Idaho, birthplace of "The Mystery Song".

Design by Jon Goodchild
Line drawings by Carol Snow
Production by David Charlsen & Others
Manufactured in the United States of America

Library of Congress Cataloging in Publication Data

Smith, Gary, 1943–
  Windsinger.

1. Smith, Gary, 1943–   I. Title.
CT275.S52424A35   978   [B] 76-21291
ISBN O-87156-192-1

Acknowledgments

Windsinger would not have been possible without the encouragement and sensitive editing provided by Cliff Cheney, particularly during its later stages when I became ill. Thanks, Cliff.

I'd also like to thank Frank Waters and Pearl Baker for encouraging me to take the first step; and my wife, Lili, and parents, Milt and Lillian Smith, who supported me along the way.

Finally, a special salute to all those I've met who know and love this land.

Grateful acknowledgment is made to Farrar, Straus & Giroux, Inc. for permission to reprint a selection from *The Selected Writings of Juan Ramon Jimenez*, translated by H. R. Hays. Copyright © 1957 by Juan Ramon Jimenez. New York: Farrar, Straus & Cudahy. Used with permission.

All photographs in this book taken by Gary M. Smith, except where otherwise credited.

FOR DICK

I am like a distracted child
Whom they drag by the hand
Through the fiesta of the world.
My eyes cling, sadly,
To things . . .
And what misery when they tear me away from them!

Jimenez

Contents

Foreword

Too many things at once, *Windsinger* is beyond my stumbling description and my small praise. It begins as a narrative and a series of lyrics, but along the way becomes the record of a life rooted in the earth—the life of a man who has spent his best years in the wilderness heartland of western America, has flown over its snowy divides, climbed its sheerest peaks, plumbed the gloom of its deepest canyons, ridden the whitewater rapids of its fiercest rivers, and retraced the trails of its prehistoric Anasazi—a man who has known intimately the still largely unknown wilderness.

As if all this were not enough, *Windsinger* reaches beyond the experiences that have highlighted a rich life to make some important comments about what I have termed spiritual ecology—the realization that the earth and all the living entities of the plant, animal, and human kingdoms comprise the one integrated life-system of our planet. *Windsinger* reminds us that each part of that system must help to maintain the life of the whole, that we injure ourselves if we do violence to any portion of the web. It teaches us that if we tap the unmolested living systems of nature, we find the ultimate source of wisdom, science, and art. In harmony, with them we can realize our true nature and function, making the connection with the deepest substratum of our essential being, our fecund unconscious. And if some of the physical frontiers which *Windsinger* explores are going, it leaves us reassured that there will always remain the spiritual frontiers beyond which we may discover the healing wilderness deep within us.

Windsinger is, like the theme of its title song, an odyssey to the Four Winds of the Mountain. It speaks of the development of awareness, the search for universality through shared experience and understanding. It appeals to us through all our key centers—our thinking, feeling and moving centers. In homey vernacular, the voice of the earthborn, it moves with range and depth through a farm outside Payette, Idaho, Utah's Robbers Roost and canyonlands, Okinawa and the Ryukyu Islands, the south and the southwest—anywhere Gary encountered himself and his songs.

I would like to say something about Gary, himself, but he speaks so clearly for himself in every line, in every song that rose out of him spontaneously from the earth itself. So rather than write a formal introduction to his multi-level book, I prefer to regard *Windsinger* as an introduction itself to an America we are on the verge of discovering as we observe its 200th anniversary.

Frank Waters

Dark Canyon, near Colorado River

Introduction

Willi Unsoeld helped his listeners find a universe under their footsteps. Instead of recounting exploits about his successful ascent of the West Ridge of Mt. Everest, he smiled across the campfire and spoke quietly of walking out of his tent at sunset, high in the Himalayas, to watch "the awesome spectacle of alpenglow wrap Mt. Everest in the celestial fire of evening." His oranges, reds, and pinks dissolved around the ice, rock, and snow in the mountains of my memory—smaller mountains than Everest, to be sure, but no less loved. How could anyone, even Willi, capture the immensity of what he had witnessed? And yet, the glory of Everest wasn't most remembered.

"When I turned to go back to the tent, I casually glanced at the frozen ground beneath my climbing boots and saw it." Willi's voice grew softer. "Nearly crushed, barely noticed, a strange companion had shared that evening with me. A tiny, beautiful white flower clung precariously to the rocks and ice next to my boot."

Willi's surprise arrival at my backcountry ranger outpost near the Robbers Roost area of Utah's desert had seemed a little incongruous to some of the campers. "Why is a mountain climber coming here?" one camper asked. I smiled to myself. Maybe Willi's questioner had missed the earlier point. Willi was patient as he explained that he was hiking with thirty of his students from Evergreen College in Olympia, Washington, into remote canyon systems like the Maze near the Land of Standing Rocks. There they planned to live simply for three weeks, studying together in a course the students had designed and named Wilderness and Consciousness.

What did they expect to learn out there in the wilderness? Willi answered our questions by describing his work in Nepal, where for several years he had worked in United States government aid projects. The poverty, disease and lack of education, he told us, were formidable obstacles for the altruistic young Americans who had come to help. In that land of scarcity, however, one thing hadn't been lacking, and it had haunted and puzzled Willi and many of his co-workers. Happiness. It was abundant.

"That's right," Willi concluded. "I decided then that maybe things should be reversed and Nepal should send their Peace Corps contingent to America to teach us how to be happy."

When I watched his students pick up their packs and begin the long hot trek toward the Maze, I knew that many of them would gain new insights about themselves along the trails. They were about to begin the adventure of finding a new love for their homeland; and perhaps after viewing the awesome wonders of canyonlands, one or two of them would look to find growing among the folds of living stone a tiny white flower.

We children of western civilization have let ourselves become dominated by an excessively rational and linear approach to life. An approach that feeds materialism, shuts out healing energy of nature and dangerously subordinates our unconscious, leaving us cut in half—incomplete. Small wonder we are swept by feelings of rootlessness and restlessness.

We have fallen into a railroad track mentality in which we believe that the consequences of our actions can be pushed somewhere far down the track of the future, and that the past has slipped too far behind us to matter. But the rhythms of the American earth work inevitably, if slowly, on this nation of restless migrants—as they worked on the early inhabitants, the Native Americans. They are transforming people like Willi's group into "native Americans," a people who are casting away their invader/exploiter approach to life, to the earth, and finding a sense of place. The cyclical nature of *Windsinger* is founded on the belief that time is a continuum and that everything is connected to everything else. We are immediately accountable for our actions. The future and the past stand at our sides, and we must act accordingly. To do so, we must discover our roots and honor the demands and limitations of the natural world.

Windsinger is a search for this world through people and music. It reflects and embodies an attempt to grow, to learn, to look deeper. It is like the Navajo Beauty Way ceremony which invites its celebrants to leave the callowness of youth to become aware of and part of the earth, establishing a mature and thoughtful relationship with her.

It begins in Beauty . . .

Gary Smith on wilderness patrol above Twin Lakes, in the Sawtooths of Idaho

BEGINNINGS

1. Color
Crayon
Morning

SAM AND PANSY

It probably all started with Granddad. I mean the summers I spent
with Sam Raby, my grandfather, on his small farm near Payette, Idaho.
Here, when I was thirteen and fourteen, something began to germinate.

Sam Raby was a farmer—a good farmer—and he planted seeds
deep. I was a wild, skinny, city kid who thought cats were to sic dogs
on, calves to ride, pigs to rope and chickens to chase. But Sam was
never one to pontificate or scold. He just went about life one step
at a time and waited for me to unwind and settle into the rhythm of
country living. His feeling for the land and quiet self-assurance
spoke for themselves.

Grandma Pansy was a different story, however. Patience with a
"foolish city slicker" was not her way. And she figured she had the
ready-made cure for what ailed me: *work, hard work,* and lots of
it. Under her watchful eye I barely even had time to torment her
obnoxious black-and-white tomcat, "Tommy." What she could see in
that cat, I'll never know. The only thing the coddled, overweight
thing had going for him was his clever "trick." A trick he had to learn to
survive.

You see, if Tommy ever set foot outside the house, one of the
scrawny barn cats would quickly whip him. So sometime in his past,
Tommy had learned to urinate down the bathtub drain. This would
tickle Grandma, making her guffaw, slap her thigh and remind
me—between gasps—not to tell company about Tommy's abilities.
(She was afraid they might be queasy about using her new inside
facilities.)

The only time Tommy ventured out was during milking and feeding time when we made our twice-daily parades to the barn. Grandma
led the way, followed by Tommy. I'd rattle along behind them with the
milk cans, keeping a watchful eye out for her mean red dog, King,
who delighted in coming out of nowhere and vanishing with a chunk
of my heel. Sometimes Grandma'd catch me placing a well-aimed
kick at my attacker and would declare it was one of those "barn cleaning
days." This didn't help my heels any, and a well-placed chunk of
rubber hose across the mutt's snout one morning sure didn't ease
the work load when she heard him yelp. However, after that, King
and I did get along.

The way to escape her work list was to volunteer to help
Grandpa. He was well into his eighties by then, and she did appreciate
my efforts to supplement his incredible pace. He was still farming about
thirty acres, feeding around twenty head of beef cattle and calves,
and helping her with milking chores. Every morning and night, seven

days a week, she'd milk the "backside" of sixteen cows while he fed "the front side."

Once when mowing the ditch with a scythe, I came off second best in a fight with two bumblebees. The infuriated dive bombers had neatly deposited their ordnance in my lip, much to the amusement of Granddad. As my lip swelled and rage abated, he reached in his mouth and pulled out his ever-present chaw of Beechnut. Tearing off a chunk of the damp tobacco, he packed it over the bee stings.

"Good for what ails you," he grinned.

"What the hell those bees go and sting me for, Grandpa?" I spluttered.

"Well, maybe you bothered 'em. All you have to do to get along is watch how you use that scythe."

By then the tobacco'd drawn out the pain. But the swelling made it hard to talk. It was a good thing too. It was time to think about what he'd said.

When the sun got hot, we'd rest under the giant North Carolina poplars in the front yard and Granddad would tell me about his adventures coming west from North Carolina in the late 1800s. He was always most fond of remembering his sheepherding days near Soda Springs, Idaho, and John Day, Oregon.

"Did you ever shoot anything, Grandpa?"

"Only if they'd bother the sheep."

"Like what! What did you shoot? Any bears, or wolves, or what . . . ?"

"Had to shoot a bear one time. He'd been killin' the sheep. So me and the dogs treed him and I shot him with my .32–20 pistol. Felt sorry about it ever since. After I skinned him, he looked just like a man hangin' there. Couldn't even stand to eat him."

Hopefully I could entice him into bringing out his prized pistol if I asked questions about shooting. It was always a high point for us grandchildren to get to look at the gunbelt and his ivory-handled Smith & Wesson with the gold stud in the grip, and hear how he won the gun.

"One time this cowboy got a

Pansy and Sam Raby

little soused over in John Day. Guess he started braggin' he was the best shot in the area, so some of the boys came out to camp and took me to town. At that time I only had an old .44 around the sheep camp, but I was used to shootin' it. At least these fellas thought I was good. So we had a contest and I won his money, saddle and horse. Finally he says, 'Well, you've won everything else. Want to try for this gun against all the rest?' I agreed and pretty soon won his gun too.

"Then I got to feelin' sorry for the fellow and gave him back his horse, saddle and half his money. I sure wouldn't give him his gun back though."

"So the big shooter rode away without a gun, huh, Grandpa?"

"No, he had a gun all right," he grinned. "I gave him my old .44."

Looking back, I know I was lucky to spend time with Sam at that stage of my life. Grandpa's aversion to needless killing was admirable.

He hadn't lost many sheep because he took care of them. Instead of waging wholesale war on predators, he quickly meted out his personal brand of justice on individual attackers, as he did with the bear. Even then he mourned because he couldn't eat it.

The wilderness had been an important teacher for Sam, and it soon would be for me. His stories helped fan my curiosity about wild lands. Sam always longed to go back.

Years later I visited Granddad when he was dying in an old folks home. While I held his hand and talked with him, he slipped in and out of the past and the present. He knew the end was near and didn't want to meet it in an antiseptic hospital bed. Suddenly I was no longer his grandson. I was a sidekick from his younger days. In a furtive whisper, he'd tell me to get the horses saddled and sneak them around to here so "we can get back to camp."

It's from these seeds that "Color Crayon Morning" has sprung. The song sketches a time when Sam rose before Pansy, at dawn, after a summer storm had begun to break up. He was struck by the beauty of the colors and tried to describe them in his own mind. Groping for words and finding his vocabulary lacking, Sam seized upon a grandchild's forgotten box of color crayons for a symbol. Even though Pansy was a real bearcat, Sam loved her greatly, as did I. He tried to articulate this love to himself, then quietly stepped outside to let her sleep a little longer. The song was my way of thanking them for starting me on a journey years ago that I've never regretted. The only thing I do regret was my failure to bring the horses around and help Sam get back to camp.

Color Crayon Morning

Just a touch of scarlet
Lighting up the rosemary;
The early dawn paints the clouds left by the summer rain.
A color crayon morning slips softly through the window glass,
Spreads across the pillow, and rests on my wife's cheek.

The cattle are a-coming slowly down along the fence row;
The mist is lifting off the meadow wet from last night's rain.
Woven through the nodding-grass, a hundred sparkling cobwebs
That the raindrops change to diamond lace in the scarlet light.

Oh, God, how I love this woman; God, I love this land;
And I'll love them, God, until I die
With the strength from these callused hands.

So I'll rise and softly close the door
Then step outside to breathe the air,
And not disturb the color crayon
        morning in her hair.

Sam Raby

GORMAN

Meeting Gorman was like being someplace you've been before and doing it all again. He looked like Granddad. My wife, Lili, and I had traveled to the quiet hollows and rolling hillsides of North Carolina, near Franklin, to find Sam Raby's roots. Sam left when he was nineteen and never returned. But his memory was still alive, and my kinfolk like Gorman Raby, Sam's nephew, made us right t'home.

Gorman's brand new pair of Red Camel overalls indicated he'd been expecting us, and his wife's welcoming dinner confirmed it. She apologized, "I don't cook like I used to what with the arthritis," then spread out fried chicken, homemade butter, pound cake, home-grown potatoes, gravy, green beans, homemade grape juice, corn bread, hot rolls, home-raised honey and fresh milk. Nope, she doesn't cook much anymore.

Gorman and I went to sit on the front porch swing to rest after supper. He talked about painting the swing like he talked about someday putting in indoor plumbing. As evening filled the valley, fireflies began to blink everywhere. We noiselessly swung on the porch and watched their lights become more intense. The whole valley filled. It seemed as if it were moving in another dimension, hurtling through universes toward some monumental discovery. I felt an urge to try to catch the moment with a time exposure, using the light of the fireflies. Then something moved out and stopped me. It came from somewhere inside. The urge left as suddenly as it had come. We were back on the swing still moving through galaxies. Gorman stirred a bit and rubbed his chin. Then he was quiet.

Next morning we learned that when Gorman wakes up and goes about his chores one of his main enjoyments is listening to his hound, Buck, running a rabbit down a far-off ridge. The music of the chase starts everything off right.

Things go by at a quiet, even pace, as if driven by a pulse many of us have ceased listening to.

Oddly, the valley's serenity and beauty are becoming a threat to people like Gorman who subsist in places like these. People from the outside with big money are trying to buy out the poorer valley people.

We met one walking along Gorman's road that day. He was wearing black dress oxfords, Bermuda shorts and sport shirt with the tails out to hide a fat gut underneath.

"What are you doin' today, mister?" Gorman drawled.

"Well, uh, well I was out looking for mushrooms," the man replied self-consciously.

"This is the wrong time of year for mushrooms," Gorman explained, adding to the man's embarrassment.

The man's red face contrasted with his white, varicose vein-riddled legs. We waved goodbye and continued in the other direction.

"Can't understand why the fellow just didn't say he was out for a walk," mused Gorman. "Mushrooms don't grow in the road. He's probably one of those Florida folk up here tryin' to buy land."

Gorman was probably right. The man's black dress oxfords weren't the best kind for tromping around in the hills. Although the man didn't know how to find mushrooms, he was someone to be feared.

I thought about Sam's old farm in Idaho. I'd never been back since it was sold to a retired couple.

Gorman Raby seemed as timeless as his valley. I guess he's always been here. Other folk left the valley to seek fortune on the outside. Gorman stayed. Now he watches as refugees from Florida and fouled nests elsewhere trickle back into the valley, hands clutching wads of money made on the outside, eagerly buying up land so they can possess the serenity found here. And they'll build cottages, subdivide the land, unplant farmers, and inflict "civilization" on the natives. Unless the valley folk like Gorman hold out.

We found ourselves hoping things wouldn't change. The place itself and the magic of the hills. The spirit of selflessness and simplicity. For if it does, if these places change like the world that swirls feverishly

Gorman Raby

around the hills, then where will we be able to go to find the simple things, the real things?

The morning light moved along the cabin wall and rested on the hunting horn, a cow's horn once used to call hounds when they were absorbed in the chase. Gorman doesn't hunt much anymore. For the most part the horn hangs silent. The chase is over. Now it only thunders through the valley to amuse a happy grandchild who listens with awe to the haunting, doleful sound.

A sound that rolls back through the valleys and hills. Stretching . . . endless.

CHARLEY

If you think you're gonna learn where Charley lived by reading this, forget it. Because, as with Sam and Gorman, to find a man like Charley, you have to take yourself back—far back—to cold spring water, hoecake, corn bread and fresh sweet cider.

I guess that's what draws me to men like Charley, Sam and Gorman. They somehow managed to slip through the clutches of modernization and maintain a fair degree of the independence of a dying era. Yet all three of these men, whose approaches to living, intangible mixes of wonder over, respect for, and closeness to their places and people, paralleled the feelings in "Color Crayon Morning," were being absorbed before my eyes by the culture they had resisted.

This fact was particularly maddening to me about the time I met Charley. I was living in the Washington, D. C., area, a situation that was more than depressing. To find the nation's seat of power, where the fate of both tame and wild lands is decided daily, perched on the banks of the open sewer of the Potomac was demoralizing. The first thing I wanted to do was to get out—go back.

A weekend escape took me to the Blue Ridge country and the home of Chuck and Nan Perdue in Rappahannock County. Intrigued with some unique cherry-wood stools in their kitchen, I asked where they came from. My hosts explained they were hand-carved by an old cooper named Charley Atkins.

"How do I find him?" I questioned. Chuck drew me a map. The map led off the asphalt, over old bridges and finally to a stream that had to be forded, unless you chose to park and hike the last mile. Charley did. So did his two sisters. He owned no car. Only went to town to get Prince Albert tobacco, and then he rode his horse.

After I had hiked the mile, the woods opened up and I walked past

Charley Atkins

split rail fences, stone walls, and sheaths of corn stalks and began the climb to the house. Charley's place was pitched on a steep flank of the Blue Ridge and was walled in on three sides by the national park that had absorbed about half of his farm. He resented having his access to the woods restricted by the park regulations, but then I wondered how any boundary could keep a man like him from his freedom.

Charley explained that he only made stools in his "piddlin' time," and didn't have any available just then. It didn't really matter. Getting to meet Charley and his sisters, Nellie and Ethel, was more important. He showed me his collection of old cooper tools and poured us both a drink of hard cider from a jug he kept hidden under some gunny sacks in the back of the cooper shack. The buzz on the cider was just right for the start of a long friendship.

After a year in the Far East, I returned to the Blue Ridge and stopped at Atkins' place. Nothing much had changed except Charley was finally getting around to making some stools for me. He'd managed to "find" some black walnut someplace and made me a matched set of five. He used an old drawing horse and draw knife, carving each stool out by hand. When he had finished, he wouldn't take more than seven dollars apiece for them. I was dumbfounded, but didn't argue. After all, what would a city feller know about the going rates for "piddlin' time"?

Obituary

Charles Clifton Atkins, 79, of Sperryville died June 12, 1973, in Newton D. Baker Hospital, Martinsburg, West Virginia. He was a retired farmer, a native of Rappahannock and the son of the late James A. Atkins and Fannie Johnson Atkins.

A graveside service conducted by Elder Clarence Frazier was held Friday in Fairview Cemetery, Culpeper.

Survivors include four sisters, Mrs. Ethel Tyner, Miss Nellie Atkins, both of Sperryville; Miss Pattie Atkins and Mrs. Lois Frazier, both of Warrenton.

Clore Funeral Home, Culpeper, was in charge of the arrangements.

—*Rappahannock News*, Thursday, June 21, 1973

Charley's Neighbor Writes:

". . . they had a sale soon after Charley's death of all his possessions. His lands and such went for pretty high prices. Nellie and Ethel had to sell the place and move to Sperryville. There were people who thought they would never leave that hollow."

2. Wind Is a River

IDAHO

Idaho was my growing place.

The nurturing I received there is something I've never taken for granted, even as a fugitive city kid rampaging among her fields, mountains, deserts and streams. Somehow, even then, before I'd grown enough to see beyond her borders and make comparisons with despoiled areas elsewhere, I intuitively felt lucky—maybe even chosen—to have been one with her at that time.

Idaho seemed immune from the deterioration around her— protected by her obscurity. She became an imaginary island of my mind, a feeling, an image to carry with me, and a metaphysical reference point. A frontier. Not in the geographical sense, but a personal, spiritual frontier, a place where the stillness of deserts can join with the energy of mountains to spawn an Ezra Pound, sustain a Vardis Fisher, and embrace an Ernest Hemingway. A place where my outdoor wanderings were punctuated by the bubblings of the unconscious—the songs that have continued to surface along my own Beauty Way.

Idaho, its name, according to legend, derived from an Indian salute to the dawn: "Behold the sun coming down the mountain!"

DAD

"You hear that sound? It's a rattlesnake. Don't forget it; he's right over there under that sagebrush."

Dad shifted me a bit on his big shoulders and continued crashing through the brush along the banks of the Little Lost River near Howe, Idaho. Occasionally he'd stop when he saw another rattler coiled in the shade under a bush, trying to escape the desert heat that had quickly sapped my own strength. He'd then reach out with his fly rod and gently flick the snake, transforming the critter into a buzzing menace. Wide-eyed, I'd scan every bush from then on from the safety of my perch, preferring not to inform Dad I was probably rested enough to make it on my own. We reached another fishing spot and he deposited me on the bank and continued my first lessons on how to catch a rainbow.

"Why didn't you kill 'em?" I asked.

"The snakes?"

"Yeh, aren't you supposed to kill rattlers?"

"Sometimes you have to, but these weren't in our way and wouldn't have bothered us. I just wanted you to learn how to see 'em, hear 'em, and stay away from 'em. After all, they have a purpose."

"Like what?" I asked.

Dad grinned as he gently placed a fly near a promising riffle. "They help keep other fishermen away from this spot."

Together we spent many years enjoying the incomparable streams and fields of Idaho, while Dad taught me the skills of hunting and fishing and gave me my first taste of freedom in the outdoors.

MOM

"It's never too early to start music lessons, so you may as well accept the fact that you're going to have them!"

I was in the second grade when my mother marched me into a music store and told the cooperative salesman she'd come to get me an instrument. Surprisingly, she left the choice up to me.

The salesman rubbed his hands together a little too eagerly and asked, "Tell me, young man, do you have any idea what you want to play?"

"Yeh, baseball."

He recovered quickly and rephrased the question.

"No, no. I mean what *instrument* would you like to play?"

"What's the loudest?"

Milt and
Lillian Smith

"The loudest? Now let me see . . . that would probably be the trumpet or the drums." Catching a look in my mother's eyes, he quickly steered me away from the drums and toward a trumpet which he blared raucously for my delight.

"I'll take it," I quickly pronounced, hoping this would dampen my mother's desire to have a schooled musician in the family. At that time, I was perfectly content to let my musical efforts be fulfilled during family songfests around Dad's piano or organ—instruments he'd taught himself to play during his youth.

She wavered only briefly, however, and the salesman was slick. He adroitly pulled out a cornet that he promised me was "almost as loud as a trumpet," then, winking at mother, "but more mellow and soothing in tone."

He made his sale; music became a part of my life.

Eventually Mom enrolled me at the studio of J.C. and Margaret Gardner in Pocatello. J.C. had once played with John Philip Sousa and was Professor Emeritus of music at Idaho State University. Margaret was a former instructor of Louie Wertz, now better known as Roger Williams. If there was anybody who could hammer out a child prodigy, it should have been these two. And to their credit, they did try, unsuccessfully.

Years later I was seduced by the guitar, and tried to joke about the transition, explaining I had a hard time singing with a cornet, but it wasn't necessary. The Gardners understood.

Equipped with basic outdoor skills, and eventually with a gentle portable companion called the guitar, it was only natural to recall Sam's memories of the wilderness and finally begin to search for my own meanings and songs in the mountains.

CLIMBING

"I hear you two are trying to kill yourselves." Blaine Gasser grinned over a well-used pipe as he watched Rich Howard and I paw over his display of climbing equipment in the now-defunct People's Store in Pocatello.

"Sounds like you've been talking to worried mothers," I replied, trying to act knowledgeable about the gear we were handling.

"I'm not saying who I talked to," Blaine teased, "but if you guys want a few pointers, I'll be happy to show you."

I never did learn who had asked Blaine to intercede, but as I look back now at some of our early climbing antics and makeshift equipment, I wonder how we survived those high-school years before Blaine put us in touch with experienced climbers, proper training and sources of good equipment. Climbers were a rare breed in those days, regarded at best as members of a lunatic fringe, so sticking together was as necessary as it was natural. Of course, climbers are always quick to argue that others in society are the real lunatics. After all, anyone who stays in the suffocating security of the flatland barrios when there are mountains to climb must be nuts.

"I wonder how many golf clubs ole' Scoop has broken in half today," chuckled Gasser as we drove past the country club on our way to the mountains.

We were going home.

For me, climbing was a love affair from the beginning. The pure physical delight made me feel like I'd only been half alive until I began. It was like slipping out of a cloud of novocaine. I began to learn about endurance and receding limits. Senses sharpened, nourished by pure air and water so cold it hurt to drink it and so clear it seemed almost tasteless. As my coordination, balance, and strength improved, it became a game to dance under a heavy pack across boulder fields, or glissade with whoops and swoops down steep snowfields. There was an essential rightness about it. . . . Joy.

Back in the city at night, I'd lay awake in amazement recalling the day's events. It was practically possible to relive every detail of a climb, so tremendous was the impact of those adventures. And in the following days it was just as amazing to watch the memories of the arduous parts of a climb dissolve like the aches in my muscles, until only the meaningful memories remained.

But probably most important, the mountains forced me to slow down. To become a better observer of my surroundings. This is one of the most commonly held joys of those who slip from the confines of the auto and take to their feet. They can see, feel and learn more in one short mile along a trail than they may ever learn while being whisked through the countryside with their noses pressed against car windows.

We'd climbed too fast and lay all spent on the top of Mt. Teewinot in the Tetons. The rapid rise in elevation had brought on the warm, nauseous sensation of mountain sickness, and I'd just ungracefully upchucked a can of Vienna sausages and was lying around trying to find my stomach. There was nothing much to do except wait, try to relax and enjoy the incredible view. My partner, John "Lefty" Reisenger, was gamboling around on the summit block and laughing about my eating habits. Reisenger once nicknamed me "the stomach that climbs like a man" because of the culinary payloads that used to mysteriously appear from my pack.

To the southwest of our position loomed the sweep of granite, ice and snow that comprises the north face of the Grand Teton and the Teton Glacier, flanked on the right by Mt. Owen. This sight is one of the most devastating in the entire Teton Range, and seeing it had been the primary reason for our scramble up Teewinot.

I wandered off to look at the flatlands, and sitting alone on the sun-warmed granite I realized that until I'd started climbing, I had always entered the hills as a predator, coming to kill or catch something to take back. Until then I'd been operating as a forager or raider. The wild-lands had been something to exploit, a commodity to utilize rather than a community to belong to. And sadly, the idle killing of many of its creatures had been for me a casual form of entertainment, not even a method of food gathering.

In my self-centered view, I had conceived of nature as existing only for my benefit, to be approached on my terms, and from which to be insulated for my comfort and security. Even in climbing I'd been guilty of a few alpine gymnastics and exploits designed to attract atten-tion and win a reputation. For the first time, I was coming to nature on her terms—not mine.

I free-climbed out on the summit block. The wind was shrieking and screaming and I laughed out loud as I climbed, bursting with the freedom of that moment. The incredible unseen and universal force of the wind was everywhere around me and I was laughing with it. John Muir had been right when he taught that "everything in the universe is hitched to everything else."

When we turned to leave the mountain that day, I took the wind along as an ally.

MOUNTAINS

The Sawtooth Mountains of Idaho birthed my first songs and gave me a chance to quit commuting to mountains and begin living with them. In 1963 the United States Forest Service opened one of its first visitor centers on the shore of spectacular Redfish Lake near Stanley, Idaho. Luckily, I was hired with the first contingent of naturalists to help initiate the new Visitor Information Program for the Sawtooths. During the four college summers I worked there as a summer ranger, my duties included conducting interpretive hikes and campfire programs, manning the visitor center, and patrolling the magnificent Sawtooth Primitive Area. They were heady exhilarating summers that provided me the chance to live and work in one of the most beautiful places on earth *and* get paid for it.

The visitors that I met in those early years continue to frequent my memory. The memories are good. I guess I've always been lucky; it's very difficult for me to remember any unpleasant or ugly personal exchange with visitors. Rather, they seem always to have reflected and given back anything positive that was brought to them, and they were anxious to learn about their surroundings. Any person who visits a natural wonder—like our parks, forests, and wilderness—and isn't somehow changed, refreshed or uplifted in some way, has been cheated. He's either been cheated by lazy, unconcerned rangers, or he's cheated himself. That's why I determined very early to shepherd visitors out the door of the visitor center and point them toward the natural delights beyond, providing them with whatever information and help they needed for a short walk away from their cars or a major trek through the back country. Visitor centers should be launching pads—vehicles of change and evolution, not destinations.

Shangri-La, somewhere in the Sawtooths

From this perspective, I quickly observed how vital our wildlands and back-country resources are in shaping positively and influencing others. Countless incidents over the years have given the lie to popular myths advanced by wilderness opponents who stereotype wilderness users as young elitists, or rich out-of-staters, who bang on tin cups, blow whistles, wear funny boots, and want selfishly to "lock out" the rest of the world from their "locked-up" wilderness. Bull.

It's impossible for me to typecast wilderness users, even after years of association with back country. I've never met one older person or cripple who would deny other citizens wild islands of hope. Rather, I've met literally hundreds of older people hiking the trails. They move a little slower but often see more. I once met a white-haired man resting against his walking stick far back in the Primitive Area.

"How old are you, sir?"

"Eighty-five goin' on ninety," he smiled. I was dumbfounded and asked how a man of his age could do what he was doing.

"Well, I guess 'a man of my age' can do what I'm doin' because I've always done it and plan to keep on doin' it."

About the biggest thing wilderness users have in common is the knowledge that backpacking can be a very inexpensive yet incredibly rewarding form of recreation. They also know the ones who are "locked out" are mostly those who have enslaved themselves to the internal combustion engine. The slaves sit popping and smoldering at the boundaries, getting soft and indolent while resenting those who are willing to try harder. All they have to do to "unlock" and enjoy the blessing within is to turn off the ignition and walk on over.

At this point I'm tempted to indulge in an esoteric discussion of the spiritual, social and biological values of wilderness—those gentle, remaining scraps of wildness that remind us of our environmental sins. However, a quiet moment on a Sawtooth trail brought the whole question out of the realm of polemics into the world of reality for me. It is the moment and reason to remember.

It had been a perfectly miserable day. A monster storm had enveloped the mountains and a heavy, cold rain was soaking me and the horses. My Stetson had long since collapsed and a continual rivulet of cold water found its way under my collar and down my backbone to puddle between my butt and the saddle. My poncho was worthless and the poor horses were enveloped in steam as they gamely struggled through the downpour. There were no signs that the storm would let up. The thought of pitching a tent in that muck wasn't too appealing, but we decided to plunge on up the trail toward Toxaway Lake. It was one of those what-in-the-hell-am-I-doing-here days.

When we neared the lake I noticed two figures approaching through the downpour. As they drew closer I could see they had ponchos draped over their backpacks and were carrying fishing rods. The smaller figure carried a large string of fish and both figures were smiling! Smiling on a day like this! What a helluva time to have to play the friendly ranger role. I was in a bad mood. We exchanged greetings.

Suddenly the father asked, "Hey, aren't you Ranger Smith?"

I said I thought I was.

"Well, you probably don't remember us, do you? I'm a policeman from Boise."

I apologized because I honestly couldn't remember him and listened as he explained how we had met three summers earlier in the visitor center at Redfish Lake.

"Well, that's understandable because you probably talk to a lot of people," the father continued. "But that's not important. The important thing is that you talked us into taking our first hike together in the mountains. We went up to the Bench Lakes and have been doing it ever since. I bought my boy and me these packs and we get out together all we can now."

For the first time all day, I began to smile. The boy proudly held up his string of fish.

"You know, Ranger Smith, I've always hoped we could meet you again so we could tell you thanks for getting us started. I really mean that. Thanks a lot . . . we'll be seein' you again . . . goodbye."

We waved again through the cold rain, which suddenly felt warmer. Climbing in the saddle, I paused in the middle of that Sawtooth trail to watch their figures dissolve through the rain, and some tears, thinking back to a father and son fishing on a Lost River years before.

THE MYSTERY SONG

While the people I met in the Sawtooths taught me far more than I was ever able to return, some of my best times were spent alone, haunting the mountains and valleys, shying away from trails. During these trips that were free from distractions, it became easier to see into myself and listen to quiet promptings that are submerged in other situations. The "Mystery Song" came that way.

I'd been reading about Chief Joseph of the Nez Perce in my little cabin below the visitor center on the north shore of Redfish Lake. The electric light drove back the darkness, and the fire in the small stove kept out a damp June cold. My own awareness of the land was deepening and it was distressing for me to learn about the barbarous actions taken against the Nez Perce, a people whose feeling for the earth was completely alien to white men like the miners and land grabbers who invaded their treaty lands and precipitated the war that ultimately destroyed the tribe, in spite of its heroic and ingenious resistance. I turned back to read again the words of the Indian prophet, Smohalla, who started the Dreamer religion.

Smohalla found the white man's mania for private ownership of his Mother Earth abhorrent. He taught that the white man's materialism or "work" distracted him so he wasn't able to "dream" or listen to the unconscious. He claimed "wisdom comes to us in dreams," and pleaded with the Nez Perce not to fall prey to the destruction and avarice the whites were bringing with them.

You ask me to plow the ground. Shall I take a knife and tear my mother's breast? Then when I die she will not take me to her bosom to rest. You ask me to dig for stone. Shall I dig under her skin for her bones? Then when I die I cannot enter her body to be born again. You ask me to cut the grass and make hay and sell it and be rich like white men. But how dare I cut off my mother's hair?

I slammed the book shut, filled my pocket with raisins and stomped out the door. It was late and there was no moon. The dew-covered willows and grass along Fishhook Creek soon soaked my trousers as I groped my way in the dark, waiting for my eyes to adjust to the clear starlight. While I was trying to negotiate a log across a quiet stretch of the creek, the stillness exploded. I teetered precariously in the center of the log, then ignominiously I lost my balance and plunged full-length into the liquid ice of Fishhook Creek.

"Sonafabitch!" My exit from the water was faster than my entrance. Shivering on the bank and trying to ward off numbness, I looked for the beaver who had smacked a warning shot on the water with his tail, causing my upset. I imagined him ensconced somewhere along the bank laughing at the awkward two-legged fool who had just baptized himself. I squished on up the creek bank and into the night. When I rounded the Redfish Lake moraine and started up the glacial valley that carries the creek to its mountain source, a down-canyon wind stepped up the staccato chattering of my teeth.

After a mile or so, I stopped to rest in the shelter of a giant Douglas fir. I got the chattering under control enough to ceremoniously chew a few soggy raisins as I began to feel the old body thermostat catch up with the cold. I sat there in the unfathomable quiet and experienced the ephemeral inner peace that the mountains offer. I found the raisins were gone and I was getting warmer. It was time to go. Leaving the tree, I found a ford and crossed the creek again, planning to cut

Gary Smith, Mount Heyburn, Idaho, by John Flannery

across a large meadow and climb to the top of the moraine.

When I reached the middle of the meadow, I stopped, suddenly feeling something alive all around me. The whole meadow erupted with movement and phenomenal energy. Shapes rose out of the darkness and thudded away. I froze, fully expecting to be trampled, but nothing happened. Then only quiet. The shapes had moved a short distance off and stopped, trying to catch my scent. The wind was blowing in the wrong direction, however, and as long as I didn't move, they couldn't tell what or where I was. Somehow, I'd managed to wander into a herd of elk, penetrating them before those downwind smelled me and stampeded. It was like standing in the center of a great natural cyclotron. Feeling, sensing, hearing, but not seeing the energy. It was another plane of reality—the place of songs.

Excited, I gained the ridge and broke out on top, cutting the trail. Maybe I was still a bit nervous but I started to whistle, and an unusual melody came as I hiked along in the darkness, a haunting tune that was filled with rich movement. Days later, I managed to work it out again on the guitar and store it away. It was the first original melody I'd ever heard. Yet, unlike the songs contained in this book, I've never been able to put words to it. Some word fragments have come, but it will be a long time before I've gained peace enough again, or have become person enough, to find the words for this mysterious offering.

Coming off the ridge, I passed the slumbering resort and walked up the road to the cabin. I entered and slipped out of my damp clothes and boots. After stuffing some wood in the stove, I dove into my cold sleeping bag, lay back and watched firelight dance on the ceiling above, projected through holes in the stove. It was time to think again of Smohalla and dreams. Then sleep.

From that walk in the Sawtooths eventually came two songs. "The Mystery Song" waits these years later, but "Wind Is A River" didn't have to wait for words. It came out quickly, words and music. Triggered by intuitions of the future, the song sounds like a lament over a broken relationship, but was actually a prediction, written long before the romance ended.

Wind Is a River

Wind in the mountains,
Wind by the shore;
Wind sings the willow,
Wind through my door.

The Wind is a River;
The earth feels its might.
My love was a spring breeze
And a joy to my life.

She was a young girl,
Lord, she was fair.
Carrying a light shadow
With snowflakes in her hair.

Features like a goddess,
With a skin smooth and clear;
Fingers light as breezes,
Wind through my hair.

Shangri-La Falls

Our love was from a distance—
Fear kept us apart;
What could we say to each other
Save that which stirred in our hearts.

Be true, be true I would pray,
So that I might know you more;
But pride, distrust, and old fears prevailed,
And now she has gone out my door.

Wind in the mountains,
Wind by the shore;
Wind sings the willow,
Wind out my door.

GROWING

3. The Ghost Herd of Spirit Lake

WHERE SONGS COME FROM

LaSalle Pocatello, grandson of Chief Pocatello, turned the rich red pipestone around in his fingers, set it aside, and went on with his story. Across the cramped room of their cabin his wife Luella continued sewing beads on a piece of soft, ermine-white buckskin she had tanned by hand weeks before. Occasionally she'd shoot a quick glance at LaSalle as he searched his memory for the exact word. He'd find it, of course; she'd smile and nod her head in agreement, and resume her work. The little stove quietly roasted everyone in the room, but Lili and I ignored the heat and strained to follow LaSalle's story.

It was a scene he had peopled before, and one that would be repeated over the years, but this day was different; I'd come to say goodbye for a while. The war in Vietnam was claiming attention worldwide, the draft was in full swing, and Uncle Sam decided I should get involved. "You can't draft me; I'm gonna enlist." So, I signed on with the Marines. Military orders had me slated to attend Defense Information School and then go on to WestPac, which for Marines at that time usually meant Vietnam. LaSalle and Luella were upset about the Vietnam situation, couldn't understand it. Neither could I.

They had a knack for adding a meaningful touch to occasions, and

LaSalle and Luella Pocatello

LaSalle's special gift that day was to share songs. He'd sent me out to the car to bring in my tape recorder and asked that the songs be kept secret until after he died. He explained that his children had been instructed to play his recordings, in his memory, a year following his death. The songs had great meaning for him.

LaSalle knew I'd been rangering in the Flaming Gorge area and had spent time in the canyonlands of Utah previous to joining the Marines. He also knew a great many of my own songs had been surfacing during that time. I was confused and couldn't really understand where they'd come from. It's impossible to know whether he was remembering these things that day, but the story he was telling, leading up to the sharing of a song, gave me a new way of understanding where my songs came from and how.

The song was about a young warrior trying to make his way back home through hostile land. It honored a natural phenomenon that had protected him. Following is a transcription of the story LaSalle told prior to sharing his secret song, a song he'd preserved long ago from the disappearing oral history of his people:

"When a young warrior was coming back home, he heard a message from Devil's Tower that gave him a warning to take off to the foothills and not be seen by the enemy. (The enemy had set an ambush for him along the normal travel route.) So he did. He took off to

the hills and traveled along unseen until he got back home.

"Now right here, we have a tower something like that in the mountains above Pocatello. My brother-in-law and I went up there one time. We heard this tower giving some kind of a sweet song. I wanted to catch it, but I couldn't do it because the wind changed directions. So I want to go back up there. If I could travel, I could go up there and could probably catch some songs. You see, that's how some songs are made. All these songs are made from a story. It's a true story that gives true songs."

When the songs were finished, all was quiet. LaSalle was no longer with us. Old dreams and memories carried him somewhere beyond the small cabin. I remembered the towers well; the towers where LaSalle had tried to *catch* a song. My friends and I had practiced rock climbing years before in the same towers. Even then, the wind had occasionally swirled through them, producing sweet sounds. However, we'd been too "busy" to really hear anything.

It was time to go. Luella handed me a beaded belt buckle. "Take this and wear it," she smiled. "We'll pray for you when we put up the lodge in the spring. You won't go to Vietnam." I told her she didn't know the Marine Corps. She just smiled back.

As it turned out, my orders were changed. First, I went back to Quantico to work on the *Marine Corps Gazette* and design a photographic museum, then I was sent to Okinawa as a public affairs officer. The Marines were withdrawing from the war. I didn't go to Vietnam.

THE GHOST HERD OF SPIRIT LAKE

While rangering for the Forest Service at Flaming Gorge National Recreation Area in northeastern Utah, I learned of an old Indian legend. Part of my duties included taking photographs and writing slide lectures for the campfire programs. My guitar was always alongside at the talks and served as an icebreaker for the crowds of campers, like in the Sawtooths before. It was always my practice to feature local folksongs if possible. It seemed to add another dimension for understanding a place.

While doing research for a campfire program, I came across a story that dealt with Spirit Lake, which is located in the Uinta Mountains west of Flaming Gorge Reservoir. According to this legend, Spirit Lake got its name from a Ute Indian tale about a brave who got lost and ended up camping on the shore of the lake. The man claimed to have seen a herd of white ghost elk swimming under the waters of the lake at night. The sound of a bell supposedly added to his terror, especially when he saw it hanging around the neck of the bull elk that led the herd. The legend says the man was so terrified by the spectacle that he lost his mind. Others must have believed this story, because, according to the legend, a superstition sprang up among the Indians which caused them to avoid the lake.

There was a special magic in the story and the lake. One night, another ranger, Richard Schreyer, and I were sitting in front of the fireplace of the old ramshackle lodge on the shore of Spirit Lake. Some people were hanging around and the lodge owners were their usual hospitable selves, encouraging us to keep on singing. The mood was mellow and a little mysterious. The only light in the place came from the fire, and it was slowly dying. Suddenly, in just the way LaSalle was to explain it two years later, I caught a melody. Just as suddenly, words started to flow. Schreyer grabbed a pencil and paper; in a matter of minutes, the complete song lay before us.

The surprise arrival of "The Ghost Herd of Spirit Lake" signalled the start of a whole new method of interpreting and understanding an area: the use of my own songs. From then on, we did the song for campers at night around the lake, and, if memory serves me correctly, one night we stationed someone with a bell behind the audience in the dark with instructions to ring it after the song was finished. He rang it all right, and damn near started a panic.

The Ghost Herd of Spirit Lake

If you hear the bell a ringin'
Before the full moon breaks,
Then you'll know that they are comin'
The Ghost Herd of Spirit Lake.

Beneath the snow-capped Uinta peaks,
There rode an Indian alone.
He came to the shore of a lonely lake
Far from his lodge at home.

Well he made his camp under sheltering pines
And built a fire so bright.
Then the fire went out with a flicker and a hiss,
And up rolled the ragged brown night.

Well the moon came creeping behind the trees,
And a cold wind started to moan.
Then he heard the sound of a ghostly bell
With a soft and muffled tone.

The bell grew louder and the wind grew still,
And a wolf screamed into the night.
He whirled and stared into the lake
And then he saw a light.

A moving herd of pale white elk
Was beneath the water's swell.
They were led in front by a mighty bull
And around his neck hung a bell.

They moved through the water without a sound
Save for a ghostly bell.
He was frozen with fear as the herd drew near
Then to the ground he fell.

As the sun hit high on Uinta peaks
He awoke from a fitful dream.
Then he fled from that place on bleeding feet
And as he ran he screamed.

When his tribesmen found him runnin' wild
He was like a man half crazed.
And to this day they skirt that place
They never go to Spirit Lake.

4. The Blue Swan

Lili and I met Lem Ward in 1970 when we were returning from watching the solar eclipse north of Kitty Hawk, North Carolina. At Chincoteague, Virginia, we had run across decoy carvers Cigar Daisy and J. Corbin Reed, and had developed a fascination for their handiwork. To call them decoy carvers is really a misnomer. Although shooting stools and decoys are still produced by many carvers, some have progressed to artistry, creating amazingly accurate replicas of various species of birdlife. Daisy and Corbin seemed to appreciate our enthusiasm and urged us to look up the deans of bird carvers, Ol' Lem and Steve Ward, in Crisfield, Maryland. We found Lem sitting in the middle of a pile of basswood chips.

"Come on in! Welcome," grinned Lem as he tipped back his hat and propped a foot up on a chair. "What can I do fer ya?"

We could tell Lem had a quick way of sizing up his visitors, almost as quick as making them feel welcome. Carvings of wild fowl lay scattered around his shop in various stages of completion. Life-sized sculptures of mallards, swans, pintails and widgeons were being shaped amid the smell of fresh wood chips and paint. Some carving tools hung on the walls, while others were scattered within easy "grabbin' distance." There was a feeling of order in the chaos of Lem's studio.

Books of poetry scattered among the wood chips and tools on the workbench revealed another side of Lem Ward. "I like to remember poems when I carve," mused Lem. "I probably once knew over 200 poems. My brother, Steve, loves poetry too, but he's on the deep side . . . stuff I don't understand." From then on, throughout our visits, Lem would always recite a favorite poem to emphasize a point. He made a lot of points; I never heard him repeat a poem.

I started wandering around the shop after a while, and suddenly jumped a little in surprise. A grouse had been standing on a stool, unseen, behind the door. Lem laughed uproariously. The grouse didn't move, but it wasn't stuffed. I moved closer and realized why Lem Ward called himself "a counterfeiter in wood." The only thing the bird lacked was the breath of life. As far as I was concerned, it was alive.

Lem apparently thought so too, for scrawled on the wood pedestal behind the bird's feet was: "This Rough Grouse to me Lemuel T. Ward is the best, most natural bird my hands has ever recomplished after fifty-two years of counterfeiting. Lem Ward 1970."

Lem was understandably proud; I knew he'd put a lot more than wood and paint into this masterpiece over the three years he'd worked on it. Its $5,000 price was designed to ward off potential buyers, to prevent him losing a special part of himself. Yet I knew that among the few thousand visitors who'd call on Lem in the next year, one of them would be a collector with five big ones in his hand.

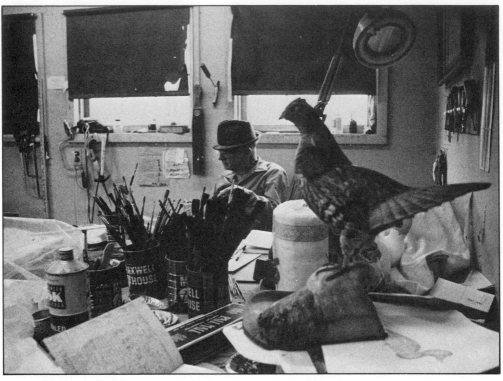

Lem Ward, his studio, and his counterfeits in wood, Crisfield, Maryland

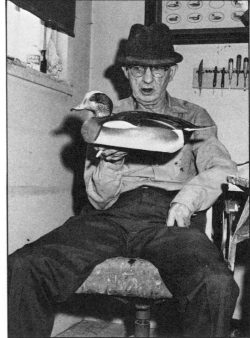

Lem was a shrewd and sensitive observer of the natural world and freely expressed alarm about the frightening changes he'd witnessed in the marshes and sounds during his lifetime. "What gets me," he said, eager to share his concern, "is the chimney swifts, the dragonflies, the fiddler crabs! You never see one! You can't find one nowhere! Used to be every blade of grass had a dragonfly on it. When you'd cut the grass, you'd cut in a cloud of 'em . . . Now they're gone. What done it? I want it explained to me!"

He talked about the disruption of waterfowl habitat, like the missile site on Wallop's Island near the Black Duck nesting grounds; and the destruction of the giant flights of redheads, "who covered the sounds and made so much racket you couldn't sleep on the boat." But mostly he wished they'd "stop gunnin'. Close the season. Doubt if it'll happen though." Lem paused, then he reached over and cradled a duck in his hands.

"You know, in spite of everything, I guess I'm the luckiest man in the world. There ain't no way to reach perfection, but you keep on tryin'. There are no two birds the same in the world. It makes a man rich to be doin' what he loves most. There just isn't anything more important than lovin' it."

We left Lem there among his wood chips and friends. I've often thought of him since and it seems to me the creations that flowed from his heart and through his hands were Lem's way of clinging to the precious birds that were disappearing during his lifetime, before his eyes.

"The Blue Swan" is probably a combination of many experiences, memories and feelings. A young boy in Idaho once shot some trumpeter swans that nested near Island Park. I can remember seeing them from the highway several times when traveling to the park. They had been regular attractions over the years for thousands of passers-by. The boy's senseless act produced moral outrage all over Idaho. Who's to say what motivated him? Dark fear of freedom and wildness lurks in us all. And yet it is a part of us too, an integral part. That's why its destruction is a form of madness. Self-destruction always is.

The Blue Swan

We watched them come flyin'
High beyond our snug town
We envied their freedom
And thrilled to their call.

We yearned for their wildness
Heard their beautiful song
When they came to live near us
Two trumpeter swans.

They lived in the marshes
Outside of our town
They built a fine nest there
And filled it with love.

They grew and they prospered
Without any cares
And ignored all the darkness
The whispers and stares.

But one day while he traveled
Far away from their home
He left her behind him
In the marsh all alone.

When one of our townsfolk
Crawled through the dark and
With a shot from his rifle
He shattered her heart.

Some people hate freedom
Some people won't love
And many fear wildness
Like the storms up above.

They dwell in their terror
While fearing the dark
And if someone is different
They tear out his heart.

At night you can hear him
Hear his lonely sad cry
As he drives back the darkness
In the empty cold night.

While many among us
Cry with his sad song
And learn how to love him
The mighty Blue Swan.

He leaves in the winter
Flies alone toward the South
Where the seasons are warmer
Than the cold bitter North.

He returns in the springtime
He returns all alone
And lives here among us
The lonely Blue Swan.

We watch him go flyin'
High beyond our snug town
We envy his freedom
And thrill to his call.

We yearn for his wildness
And join his sad song
We share all his love now
The love of Blue Swan.

5. The Ballad
of the
Golden Eagle

Some of the boaters who frequented Flaming Gorge during the time I worked there had to be the strangest amphibians on earth. Observing this species at close range was easy because the Forest Service would rotate us among the boat ramps as part of our duties. While there, we were supposed to show the flag and collect launching fees. At fifty cents a boat, we would have to launch between forty and fifty boats a day just to recover what our salaries cost. Can't ever remember launching that many boats in a day. Maybe they felt we needed a little comic relief from the mob scene we'd usually encounter in the visitor center at the dam.

Each morning at Cedar Springs Boat Ramp we'd sit at the ticket dispenser watching boats drag their owners down the ramp and out around the lake. After the early birds got their boats launched and on their way, things would slow to a sleepy crawl and we'd drowse in the warm morning sun while waiting for the day to officially begin. That happened when "Old Hemingway" stuck his head out of his camper door to hack the nicotine and phlegm out of his lungs.

"Hemingway's" real name was Frank Gridley, and he was probably the most loved man on the lake. I loved him because he'd been a squatter in the Cedar Springs boat-trailer parking lot for several summers. To torment the officious bureaucrats in the recreation area, Frank would park his tote-goat by the boat dock when he left in the boat he always illegally tied up there. Supposedly the dock was to be used only by those who were launching or loading their boats. Most of us ignored the regs in Frank's case, though, and would quietly "persuade" any co-worker who bugged Frank about the rules to knock it off. Mornings just started off right when old Frank was around, and we were determined to keep him there—a quiet salute to nonconformity.

After Frank got his heart started and his smelly little black-and-white cocker spaniel, Freckles, out the door, he'd climb on his tote-goat and drape a fishing net loaded with cans of beer over the handle bars. Then, with Freckles leading the way, he'd clatter and roar by the ticket dispenser and down to the dock.

Actually, both Freckles and Frank smelled a bit; they lived almost entirely on the fish Frank never failed to catch in the reservoir. After Freckles made her daily attempt to run down a golden-mantled ground squirrel that lived along the edge of the ramp, they'd both climb into Frank's crusty little motor boat and putt across the water. If it rained, he'd keep on fishing, stopping just long enough to stick an open umbrella's handle down the back of his neck. Around noon they'd come back and return to the camper. I used to sit in his doorway sometimes, fight back the fish smell and try to eat a sandwich.

Hopefully Frank would spin one of his stories before his afternoon siesta.

"Frank, how would you like me to con some of the girls up at the Lodge into comin' down to clean out this boar's nest?" I joked one day.

"Oh my no . . . hell no! Why it took me years to get it the way I like it. Good gawd, I retired so's to get away from worryin' about things like that," Frank would grinningly protest.

I'd look at the layers of fish blood and crud on the floor and say, "OK Frank, have it your way."

One time I asked Frank how he had made his fortune, enough to retire in "style" the way he had.

"Oh my yes," he explained, "I'm well enough off all right. My family made its fortune sellin' whiskey to the Mormons in Salt Lake."

"That sounds more dangerous than sellin' it to the Indians, Frank. How'd you pull it off?"

Old Frank's eyes would get bigger behind the thick lenses of his glasses as he'd chuckle, "Wasn't dangerous at all. The Mormons loved our product. Why, every Saturday night the buggies or cars of a lot of the higher-ups would line up behind our back door waitin' for their supply."

"Are you puttin' me on?"

"Heavens no. Why we had a specialty brand they all seemed to really crave. We called it Eight-Year-Old Kentucky Dew."

"You mean you made your own whiskey in Salt Lake?"

"Well, not exactly. You see, we would take a four-year-old barrel of whiskey, a four-year-old barrel of Kentucky bourbon and a barrel of water. Then we'd mix 'em all together and stir a little molasses in it for flavoring. Four plus four makes eight. Eight-Year-Old Kentucky Dew! They bought it as fast as we could mix it." Frank drew contentedly on his cigarette and lay back for his afternoon nap. It

"Hemingway"
Frank Gridley

was time for me to go back to guard the ticket dispenser.

"Hey, Frank," I called from the door, "didn't you ever feel a little guilty about cuttin' your licker? You probably turned a pretty tidy little profit, didn't you?"

Frank cackled a little under the blue cloud of smoke over his bed. "Well now, the Good Book says we all gotta pay for our sins. We just helped out the Lord a little bit, and let the drinkers pay us."

When things got too uproarious at the docks, it was usually time to go shake down the campers. A new law had been passed by Congress called the Land and Water Conservation Fund. It required the collection of some fees from those who used improved recreation facilities in federal areas. The monies were put into a national fund that could be used in part by state and local governments for the development of recreation projects. It sounded like a good idea, and we usually had little trouble getting people to cooperate. However, the old conscience problems rose again. It always seemed like it was costing more to administer the collection of fees and pay our salaries than we ever collected.

We always timed collection schedules to coincide with dinner time when the campers were more available for fleecing. We'd wander up to a camp looking gaunt, overworked and thirsty. A few longing looks at the cookstove usually elicited the following type of conversation.

"I sure hate to interrupt you folks so close to dinner time, but the Golden Eagle's come to take another bite from you overburdened taxpayers."

"That's OK. We already have our passport. Does that mean we still have to pay?"

"Oh no, sir, if you have the passport you're covered."

"Say, ranger, why don't you stay and join us for dinner?"

"Well now, sir, I sure would hate to impose, and besides, I have to cook for my two roommates tonight. We take turns, you know."

"Well, heck, bring 'em along. There's plenty for everybody."

"Well, if you insist. I'll turn in the truck, get 'em, and hurry right back!" I'd be off in a flash to round up my waiting roommates, who were always eager for a handout.

We always tried to give back more than we ate—fishing tips, stories about the area, a little guitar music and singing after dinner. Hopefully, those we mooched from enjoyed our meetings as much as we did. One song they seemed to particularly like, especially if we'd just collected from them, was "The Ballad of the Golden Eagle." It came one night when I was driving down to a joint Park Service and

Gary Smith, ranger, canyonlands

Forest Service picnic. There was a friendly rivalry between the two agencies at that time, particularly since they both managed Flaming Gorge until the politicians decided to turn it all over to the Forest Service. The words will go with about any stock blue grass melody.

The Ballad of the Golden Eagle

I am the Golden Eagle, better known as the Claw!
I got my job from Uncle through the Land and Water Law!
So if you see me comin', boys, hold the Claw up high;
And make the tourists pay a buck to help the Eagle fly!

A Lady Bird said to me, "Come now, let us fly.
We'll tear the nasty billboards down and burn 'em to the sky!
Then we'll clean up all the litter, bottles and tin cans;
Load 'em on an aeroplane and drop 'em on Viet Nam!"

Old Pete Santee said, "Listen, Boys, this will not be a lark.
For if you cannot make 'em pay, we'll ship you to the park!"
Now all us Forest workers know this mighty well:
That bein' sent to a national park's like goin' straight to hell!

So the Claw will grab and rip and squeeze to make the tourists
            pay;
So they will have a nice clean place in which to recreate.
They'll pay their seven bucks to have their fun and games,
Then they'll scatter round their bottles, cups, tin cans and paper
            plates!

6. To Marsha Wherever She May Wander

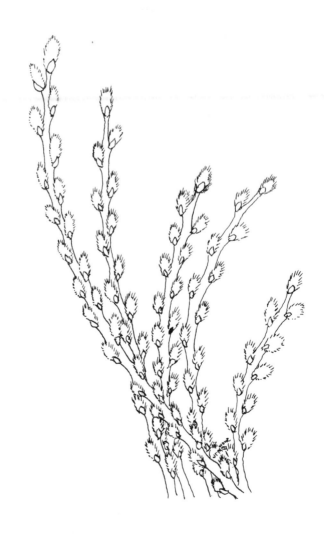

However fine the mooched meals in the campground were, the cuisine served up in the sheep camps around Flaming Gorge really tempted palates and exploded waistlines, particularly the camps run by Basque herders like José Ezcurra and Ramon Lastiri. It was common procedure for us to grab a couple of jugs of wine, round up some girls from Flaming Gorge Lodge and head out for an evening of feasting, singing and dancing around the sheepherders' campfires.

I'd been introduced to the Basques of Idaho years before by a folk-singing priest named Ramon Echeverria, a hearty epicure who would have passed for Friar Tuck. Father Echeverria would sit down to jam sometimes, and I'd listen in amazement as he'd sing for two hours and rarely repeat a language. He could sing in about thirty different languages, five or six of which he also spoke. The priest taught me to play some Basque songs and dances on the guitar and later introduced me to the colorfully exuberant Oinkari Basque Dancers of Boise, Idaho.

The Oinkaris are sponsored by the Basque community of the Boise area and have played an important role in preserving Basque culture in Idaho, while serving as traveling ambassadors of good will. The vitality of the Oinkari Dancers is marvelous to watch—or, even better, to join, particularly at the annual Basque festivals held in Boise and Elko, Nevada. In Boise they have traditionally centered around the Sheepherder's Ball, when the shepherds leave the hills and flocks long enough to let off pent-up steam and renew old acquaintances.

The festivals generally feature strenuous athletic contests adapted from everyday work activities. Weight-lifting contests, for example, use heavy, concrete-filled iron cylinders that weigh either 250 or 304 pounds. A contestant is scored on the number of times he can jerk the weight from the ground, into his mid-section and roll it up onto his shoulder within specified time limits. It closely resembles lifting large sacks of grain. I once saw a contestant hoist the 304-pound cylinder to his shoulder fifty-seven times in ten minutes!

A sport that my sheepherder friend, José Ezcurra, seemed to particularly relish was wood chopping. Not something for the squeamish to try. A contestant uses a large, razor-sharp Basque axe. He stands on top of the log and swings downward between his feet, thereby getting the strength of his leg and back muscles into the blows. Logs are often over fifty inches in diameter.

There's a possibility that the double-bitted axes we see today might be related to an earlier Basque invention, the double-headed axes that Hannibal acquired for camp tools during his march through the Pyrenees on his way to Rome's back door. A master diplomat, Hannibal signed a peace treaty with the fierce Basque tribes before they guided his multinational polyglot army safely through their

Basque sheepherders

mountain lands and the Alps around 47 B.C. Even then the Basques were a self-contained and democratic people whose roots and language mystified invaders. Today some linguists and anthropologists consider the Basques a people of unknown origins, whose language is most likely rooted in the Stone Age. Sharp tools and instruments are derived from "aitz," the Basque word for stone. The name for axe is "aitz-kor."

Although José lacked the traditional Basque axe at the sheep camp, he was always eager to demonstrate his technique with an old single bit. I once made the mistake of telling him I'd like to get some pictures of him banging away on a log. When I returned a day later, José had gone overboard. The little rascal had felled a large, healthy pine and had a series of notches chopped all the way up the trunk. Generosity and a flair for showmanship are Basque qualities not limited to the dance floor.

Although Basques are well known throughout the American West as skilled shepherds of high integrity, most entered this lonely occupation because of the severe language barrier they faced.

Historically they were best known for their seafaring skills. Supposedly Basque fishermen had been fishing for cod off the coast of Newfoundland before Columbus set sail in the *Santa Maria* with a Basque navigator, Lakotza, and Basque crew.

They have gained a reputation for going out of their way to have a good time, especially if there is singing and dancing involved. Probably their most joyous expression of unity and love for the earth is found in their songs and dances, whose origins stretch far back to mysterious Basque beginnings. Hannibal was reportedly honored by some Iberian recruits with the impressive Ezpatadantza, a sword dance that resembles one performed today by the Oinkari Dancers. A Roman geographer named Strabo described Basque dances around the time of Christ where the whirling dancers were accompanied by the txistu and drum. These traditional instruments are a recorder-like flute played vertically with one hand, and a small drum played simultaneously with the other hand.

Once in the Sawtooths while giving a campfire program, I looked up to see a good-natured grin plastered all over the rosy, cherubic face of Ray Mansisidor of the Oinkari Dancers. With him was Al Erquiaga from Boise and John Barrutia from Mountain Home. Ray had his street accordion, John had his pockets full of harmonicas, and Al was rarin' to dance. Impulsively, they'd jumped in a car at Boise and headed for Redfish Lake in the Sawtooths earlier that day. Now they stood facing me behind the unsuspecting audience. I grabbed the guitar and gave them a quick nod. Suddenly, whirling, whooping and singing

Oinkari Dancers, Boise, Idaho

Basques erupted among the campers, who quickly caught the infection, forgot their own inhibitions and joined in.

It takes a great deal of endurance and agility to keep up with Basque dances. In fact, one might describe them as musical gymnastics, and the Oinkaris perform them all with great artistry and color. But one really feels how close to nature these ceremonies are after a fine meal of fresh lamb, sheepherder bread, red wine, and rich, black coffee.

Marsha often joined in the festivities of the Basques around Flaming Gorge, and was a thoughtful, sensitive companion on other forays around the countryside. Our first outing had begun before dawn once when we traveled to the edge of some cliffs overlooking a place called Red Canyon and the Green River to watch the sunrise. While we were returning, the morning light made its way across and through regal stands of ponderosa pines . . . "like a golden-crystal dawn." Word fragments came at that time, and Marsha's song grew around them during the summer, especially during times spent around the campfire with our Basque friends and some of her co-workers from Flaming Gorge Lodge.

There in the firelight, with the full moon frosting the meadows near Spirit Lake, José would rise with uplifted arms and spin through the intricate steps of the jota, a blissful, carefree pronouncement of "I am." Ramon would sing and I'd play the guitar. Marsha and other girls would try to follow José's lead, all humming and singing the melody.

We were the dancers Strabo described 2,000 years before who danced separately yet together, this time without the flute and drum. The sheep regarded us curiously; the dogs slept under the wagon.

If one could have pulled back and risen to the clouds that raced overhead, playing tag with the moon, and looked down from that vantage point at the tiny orange sea of light in the darkness, he might have glimpsed the whirling figures, fragile specks of life spinning down through time, with upstretched arms like tree branches, joyously linking earth with the rest of the universe.

The Oak of Guernica can never die.

To Marsha Wherever She May Wander

To Marsha, wherever she may wander
May her love flow down like a golden crystal dawn.
Will she think of me
When the mist rolls through the willow tree?
And may the angels watch, and touch, and kiss,
And keep her loving eyes for me.

We rode out on a golden Saturday morning;
From the edge of the cliff we watched the Green River a flowin'.
The dawn was real and fresh and filled with the summer's joy.
Then the wind began to softly sing this song for the girl from the boy.

Love must be strong as thunder and gentle as a light spring rain;
Each day it must renew itself and begin to give again.
This message echoed in our hearts, as we roamed through mountain
      glens;
I reached out and touched her eyes, then I began to give.

And so we'll rise up at dawn and give thanks for another day;
A day in which to love again and give ourselves away.
And as I walk life's winding road and follow a thousand tracks;
A quiet voice within my soul says you'll be going, going back.

To Marsha, wherever she may wander;
May her love flow down like a golden crystal dawn.
Will she think of me
When the mist rolls through the willow tree?
And may the angels watch, and touch, and kiss,
And keep her loving eyes for me.

7. A Mist of Blue

An old range management professor at Utah State University used to jokingly call Utah's desert areas "Ten-Eighty Range Land, because a cow has to have a mouth ten feet wide and go eighty miles an hour to get enough to eat."

There's a lot of truth in the epithet. The lack of water in the American desert calls the shots there. Great portions of the land are suited only for grazing; crop-land agriculture is, for the most part, completely out of the question. Those folks who haven't realized that this country is drastically overpopulated now, and look to those "wide open western spaces" to sop up the extra humans, had better tell the youngsters headed this way to bring their canteens—full.

Old Joe Biddlecome learned to live with the desert in southern Utah on his Robbers Roost Ranch, west of the confluence of the Green and Colorado Rivers, above the Orange Cliffs. Old Joe drifted onto the Roost in 1909 after he had been "invited" to leave Colorado. Apparently Joe's cattle herd had an annoying habit of growing faster than other folks', so his fellow ranchers decided they could no longer afford him as a neighbor.

"Some people claimed he did some rustlin'," explained daughter Pearl Biddlecome Baker one day, "but hell, everybody did. He was just better at it than most."

Pearl laughed heartily as we jounced across the Roost in the Park Service power wagon. As miles of her father's former cattle empire slipped past the truck windows, Pearl reminisced about her life on the Roost and some of the characters that had frequented it.

She matter-of-factly described the labor and sweat her family poured into the corrals that are still used today on Twin Corral Flat. The cowboys had cut and hauled juniper logs and woven them into towering, circular walls that remain impervious to weather and range bulls. A gate between the circles lets the cowboys sort the herd, cutting out calves to be branded, vaccinated, dehorned and castrated. When I remarked on the corrals' design and durability, Pearl laughed, "My father was building his empire . . . an empire he wanted to last." She pointed to a semicircular wing of juniper fence that extended from the main gate. "When the cowboys rounded up the cattle and got 'em headed this way, the herd would hit that fence and follow it on around until, wham, we had the gate closed behind 'em.'

Pearl was sixty-eight years old when we traveled the Roost together in 1973. Yet she carried her years with the tough resiliency of the weathered juniper logs she quietly reached out to touch while she continued her reflections.

According to her, Old Joe probably would have been perfectly contented to live the life of a nomad, following the herds and living out of

Warbonnet Peak,
viewed from the
summit of Pack Rat
Peak in the Saw-
tooth Wilderness
Area.

John "Lefty"
Reisenger on the
summit of
Mt. Teewinot in the
Tetons.

LaSalle and Luella
Pocatello.

José Ezcurra, near Utah's Spirit Lake.
Basque sheepherder Ramon Lastiri, near
Spirit Lake.

Pearl Baker at Joe Biddlecome's corral.
Art Ekker at Robbers Roost corral.
Along the Orange Cliffs.

Hedgehog cactus blooms in Maze.
Mission District, San Francisco.
A. C. Ekker exercising.

Joe and Millie Biddlecome, courtesy of Pearl Baker

Dutch ovens and on the ground, something their family had done for years.

But Joe had two daughters and a wife to think of, so at the persistent urging of his wife, Millie, he reluctantly agreed to build a cabin near Blue John Spring—named for the outlaw who ferried supplies from Hanksville to the base camp of the Wild Bunch in Upper Pasture, between Spur Fork and Horseshoe Canyon.

Millie protested and said she would wait until Joe found her "a place with a view," a wish that was granted later after Joe jumped a black wild horse pawing at a small seep in a draw. He roped and tamed the horse, named it "Crow," and built his juniper log "castle" near the spring, now appropriately called "Crow Seep."

At last Millie had her view, a magnificent sweep falling away toward the Henry Mountains over the breaks of the Dirty Devil River. And Joe had his headquarters for what would later be known as the Robbers Roost Ranch, so named because the trails and camps of infamous characters like Butch Cassidy, the Sundance Kid, Elzy Lay, Silvertip, Blue John, Flat Nose George, and Indian Ed Newcomb were still relatively fresh when the Biddlecome Bunch moved onto the Roost. Silvertip Spring and Bad Man Trail are just a few of many landmarks around the Roost that mark the passage of the outlaws (outlaws Pearl eventually described in her book, *The Wild Bunch at Robbers Roost*).

Old Joe had left many landmarks in his wake, too. Using a series of strategically-located juniper and pinyon log corrals, existing networks of horsetrails —outlaw and otherwise— juniper drift fences across canyons, and developed springs, he interlaced sections of his range land so he could efficiently ride herd on his dispersed cattle.

Joe and Millie Biddlecome, courtesy of Pearl Baker

Everywhere I roamed on the plateaus and in the canyons, I'd find examples of his energy and intelligence. I'd often find trails blasted and carved by hand down slickrock walls and ledges to springs nestled in alcoves below. Naturally, Joe had developed the springs with durable fir troughs, probably named them, and more importantly, filed a claim on the water, effectively enabling him to control large sections of public range land without further burdens of ownership. Burrow Seep, Outlaw Spring, Roost Spring, Trail Spring, Crow Seep, Frenchy's Spring and others sprinkled through the countryside allowed Joe to utilize the canyons and plateaus for his far-flung herds.

Joe only homesteaded around Crow Seep, keeping his operation simple. When old enough, Pearl and her sister Hazel joined Joe and Millie to serve as extra hands, keeping overhead to a minimum. The herds prospered. The empire became a reality.

After Joe died in 1928, newlywed Pearl and her first husband, Mel Marsing, managed the ranch on Robbers Roost together for a little over a year. The young couple was bound together by the expanse of their isolated realm and their own happiness. Then, tragically, Mel was injured while working livestock, and he died in 1929. Pearl fought sorrow by turning to the work at hand and managed the ranch for the next five years, finally selling the operation to her sister's husband, Art Ekker, and keeping the Roost "in the family."

Listening to Pearl's description of her young love forty-four years before called to mind thoughts of "A Mist of Blue," a song I wrote for a friend who traveled many ridges in northern Utah on horseback with me. Like Pearl and Mel, and the others described in this chapter, my companion's love for the freedoms of the West captured my admiration.

A Mist of Blue

Come close to me, take my hand, watch my eyes
Don't leave so soon, hear my soul—how it's crying.
I've searched for you, through a hundred wooded hills,
On a thousand mountain tops, in a million emerald glens.
And now I find you here standing by me on a windy ridge;
Come walk with me, take my hand.

Don't change your loving ways no matter what the world may do.
Let the wind comb your hair, may you always seek the truth.
Although the world revolves through a hundred thousand strifes,
We shall walk together, and together chase the light.
For now I find you here, standing by me on a windy ridge;
Come walk with me, take my hand.

So we shall climb together; let us soar above the heights.
May we see with clear perspective and try to live what's right.
For I'll love you with the softness of the mist all wrapped in blue,
With the strength of granite mountains, all these things I promise you.
For now I find you here, standing by me on a windy ridge;
Come walk with me, take my hand.

Pearl Baker

8. The Cuttin' Horse

Art, A.C., and the rest of the Ekker family have continued to improve the ranch at Robbers Roost, while preserving warmth and informality in their operation. They are proof positive that the Old West is still alive and young. Years ago Art and A.C. also started a guide service called Outlaw Trails, running jeep and pack trips into what is now sometimes loosely referred to as "western canyonlands." Eventually, A.C.—who is an excellent white-water boatman—expanded the operation to include float trips down the Green and Colorado Rivers.

I'd heard about some of A.C.'s escapades years before in the Sawtooths from his old rodeo partner, R.J. Smith, of Salmon, Idaho, who worked summers for the Idaho Fish and Game Department. Both R.J. and A.C. rode for the University of Utah Intercollegiate Rodeo Team. In fact, A.C. was so proficient he eventually walked off with the All-Around Cowboy title for the National Intercollegiate Rodeo Association in 1967. But then A.C. had a little edge on the competition. After all, he was an honest-to-god cowboy, born and raised in the remote haunts of desperadoes like Butch Cassidy and the Wild Bunch. Plus he'd been schooled by his father Art, and there aren't many finer hands around.

It's easy to understand why Art and A.C. love this country as you roam across the vast, rolling green parks of the Roost, Twin Corral Flats or the high part of the Spur, and gaze off to find the Henry, LaSal, or Abajo Mountains rising through blue distances. The mountains loom like drummers beyond the edges of the plateau, pulsing their energy through the earth, up through your feet and into your heart.

You become an expression of the place. It seizes you, seduces you, transforms and heals—all at the same time, before you even know what has happened. You want to absorb it and be absorbed by it, to possess its freedoms, to feel its rhythms. You want to be its lover. The pronghorns share your ecstasy as they whirl over the land's folds; the horned larks play tag with your passage. You wait for unfenced range cattle to move drowsily out of your path; Ekker's remuda of horses runs before a thunderstorm rolling in from the southwest. Wild burrows scamper through yellow sunflowers and beeweed back to safety in Spur Fork. The mule deer watch.

The rhythms hold you and invite you to step to their beat for a while. To slow down. To linger.

I admired A.C. and Art Ekker for their frank, unabashed love for the land. I knew they would take care of it.

It's a local tradition for friends and family members to rendezvous at the Ekker ranch during roundup and branding time, a festivity I joined in 1974. After the cattle were gathered and corralled, the new calves and unbranded cattle were quickly separated from the rest of the herd by riders on cutting horses. Branding and dehorning irons heated in a pinyon fire, with dirt piled over the shafts to keep the handles cool. When the irons got hot, the action began.

A.C.'s sister, Gay, roped a calf in the "bunch" and dragged it behind her horse, kicking and bawling toward the fire. On the ground, I slid along the lariat to the calf, grabbed the critter and "flanked" it, throwing it down. After releasing the noose around the calf's neck, I hog-tied it and left it to await the iron. Gay dragged out another calf.

Art or A.C. then approached with a hot iron and seared the brand into the animal's hide. A few quick flicks of a pocket knife transformed a bull calf into a steer, and the "Rocky Mountain oysters" were laid along corral poles for next morning's breakfast. The young nubbins of horns were dug out of the animal's head with a twisting motion of the knife, and a hot iron was used to cauterize the wounds and stop bleeding. The calf was then vaccinated, tallied and released to stumble, stunned and bloody, back to the bunch.

The corral was transformed into a dustbowl filled with weary, cursing men struggling with frightened, half-crazed animals bawling and rolling their eyes. Dudes stumbled around, ankle-deep in dust, trying to keep pace with Art and A.C. who patiently tolerated their inexperienced "hands."

"Bra-a-a-at!" I suddenly found myself under a hefty calf that out-

Robbers Roost corral

wrassled me. I cursed, spit out the cow shit and dirt and grabbed for the flank again. Art shook his head, fought back a grin, and stepped to the water jug. The wild cows and older calves that had escaped earlier roundups were then team-roped. A.C. snared a head; Art heeled the hind feet. They quickly stretched her and the flankers stepped in to knock her on her side. The hot iron sizzled, staking its claim.

Heat, sweat and blood. The fine dust rose like a curtain, blinding and choking animals and men. Cowboys went through the motions, weary repetition. The smells of smoke, burnt hair and flesh interwove the players and actors. The scene blurred, and was over.

Like the Ekkers, Jim Stevens and his father, Willis, were Utah cattlemen. Their ranch was located in an extremely rugged and remote area of Utah, forty-five miles south of the tiny town of Ouray on Willow Creek, near the eastern edge of the Uintah and Ouray Indian Reservation. If the roads were good, you might be able to drive from Vernal, Utah, where the Stevens' kids had attended school, in three to four hours. A sudden thunder shower, however, would put you axle-deep in mud and clay. Not much you could do then except throw watches and schedules out the window, settle back with your thoughts and wait for the road to dry out. Some might feel the Stevens weren't located "near anything." Willis knew better. He was near everything . . . everything important to him, that is. The desert was his home. He'd been ranching in the area since he was eight years old when his family had settled on nearby Hill Creek. He

Art Ekker, Robbers Roost corral

remembers how they had to struggle to earn a living off the land. As a boy, one of his duties was to sleep in the grain fields at night to protect their crops from the wild horses that roam freely there even today.

That's what first brought me to Jim's ranch . . .

"Do you think we could catch one?" I asked while munching on a sliced tongue sandwich Jim's mother, Jean, had fixed for our ride. We were sitting on the edge of a cliff watching a small, unsuspecting band of wild horses in a grassy valley below.

"We might," grinned Jim. "What're you gonna do with it?"

"Probably let it go. Hell, I don't know what I want to catch one for; maybe just to see if we can."

We split up and rode into position. Jim, with the fastest horse, would wait at the head of the valley, while I slipped in below to start the wild horses in his direction. My horse was a stocky, brown cow pony called Ron who seemed to sense what we were about. I rode low on his right side and gave him his head. His ears strained toward the herd as we approached at an easy walk. I could feel his whole body tense under me. We drew closer.

Suddenly, an old mare saw me trying to stay low and snorted a warning. They were gone, and so were we! Ron came uncoiled. Somehow I managed to hang on. I shook out a lariat, even though I didn't know how to use it very well, settled into the breakneck chase and relished every gut-gripping moment of it.

I held Ron back a little while to let him warm up slowly. We still had a good distance to cover before Jim would get his chance. Ron seemed to resent my intrusion. Every fiber of his body strained to close the gap; to become one with the force we chased before us. I felt sorry for him, burdened as he was. When he got his wind I turned the rest over to him. The harder he tried to catch his free brothers, however, the more they pulled away.

The valley became a blur, a whirl, a longing.

I pulled Ron back some as the trail got steeper, to keep him from getting wind-broken. He fought it stubbornly for a while, then acquiesced.

Soon Jim sprang from ambush on his horse and flung his lariat. He missed; they whirled out of sight; I breathed a sigh of relief.

"Sorry about that," grinned Jim. "Rope snagged the brush just as I threw. It should've been easy." We coiled our ropes, fastened them to the saddles and started the long ride home.

I don't know if I ever told Jim that day, but I was glad he had missed.

"I don't mind eatin' mutton at all," laughed Willis, as he dove headfirst into the 'table flock' he kept around the ranch house. The sheep scattered and Willis struggled to his feet, holding the hind legs of a terrified young ewe. "After all, that's one way us cattlemen can get 'em off the range."

Jim stepped up to help. In a matter of minutes they slit its throat, cleaned and skinned it. Some city girls who had traveled out with me for a visit watched open-mouthed from the sidelines. They were learning about the basics.

After the sheep was hung in the shade to cool, we stomped into Jean's snug kitchen to dig into one of her good ranch dinners.

We washed up in water dipped from the creek. Drinking water, hauled from a mountain spring on their summer range, was drawn from a storage cistern outside. A propane refrigerator stood on one side of the kitchen; gas lamps provided light. Telephones and electricity hadn't intruded on this small corner of the planet yet.

Jean had wisely decided not to serve mutton that night. Instead, she plopped down a plate full of delicious, aromatic venison in front of the girls. Willis grinned. Dinner was served.

After the meal I sat on the front porch and watched Jim and Willis braid rawhide hackamores. The two men talked quietly about the next day's work; women's voices and the rattle of dishes filtered out of the kitchen. The men were resting while their rough hands passed the time creatively. Tomorrow would be a time to begin again to begin. Willis would resume his struggle and passion with the desert. Six hundred head of cattle depended on him.

Looking back, I suppose Willis knew the desert would win in the end. Jim and other sons would go their way; he and Jean would grow tired and have to sell the ranch. Still, he wouldn't be able to stay away. He'd take up trapping coyotes and keep on returning to the desert.

But that was all years away in the future.

Tonight, they were a father and son speaking softly together in the growing darkness.

"The Cuttin' Horse" was caught in honor of cowboy Jim Stevens' birthday, and its rhythms are those I felt while riding horseback over the Stevens' ranch. It's offered here with respect to all conscientious stockmen, like the Ekkers and Stevens, who treat the land as a community, not a commodity.

The Cuttin' Horse

Hold now; cut now; stay onto that stray, old boy!
Crouch now; jump now, slide into that turn!
Bluff him; run him; head him back to the bunchin' ground—
Go, you cuttin' horse!

He could run like the wind over a hundred hills,
And the thunder roared his name.
There was none could ride,
There was none could slide,
Like my cuttin' horse.

As far as his looks, they didn't show much,
And his stock had won no fame,
But his heart was strong;
You could work all day long,
On my cuttin' horse.

A steer broke clear from the hold-up man.
My horse was right up on his hocks,
The rope snaked out,
And the loop was set;
My God, how that horse could stop!

MOVING

9. Something Big

CONEY ISLAND

New York cab drivers are some of the finest characters in the world, at least if most are anything like the ones who got me to and from the airport. My big-city prejudices were running rampant after I settled through the dirty brown air at the airport. The Hong Kong flu didn't help. In short, seeing the Statue of Liberty up to her armpits in smog hadn't been a good way to start. The hack driver helped turn things around. He didn't fit the mold of what I'd been warned to expect.

"Where you come from? Out West?"

"Yeh, Idaho. How'd you guess that?"

"Can't get much farther east than New York," he laughed. "But, I'm pretty good at guessin' after drivin' cabs so many years."

He was a good driver, and after he learned it was my first trip, he kept up a running commentary about the points of interest we passed on the way to the hotel. The man really dug driving cabs. Most of all, he loved the city, and he was as anxious to share her as I would have been to share an Idaho wilderness with him. Told him as much, but he was content to stay. He talked about getting his suntans through the car window by keeping his shirt open in the summer.

He wanted to know what a "wilderness lover is doin' comin' to the city?" I told him about the environmental documentary I was making with ABC. Told him about song writing and how ABC had bought some of my songs for the soundtrack. One song even seemed to have the beat of his hack.

"What's it called?"

"Something Big."

"I'll be sure to watch for it."

I don't know if he ever saw it but I hope so. I always wanted to give something back to that driver. He even refused a tip, which I guess, by New York standards, is a real mind-blower. I insisted; so did he. He won. I wanted to know why.

"Cause I want you to save some of that wilderness for me. Besides, you guys might help get the air cleaned up. It even gets to me sometimes." He stopped at the motel. I got out and turned to watch this destroyer of myths gallop away in the Manhattan traffic in his four-wheeled charger.

Next morning found Mary Margaret Goodwin of St. Thomas, Harold Haskins of Philadelphia, and me traveling around the city with some P.R. men to shoot publicity stuff for the program. Mary Margaret was being featured for her work in marine biology; Harold was a social worker at Temple University who had helped turn the Twelfth-and-Oxford street gang into a business corporation. We represented diverse environmental problem areas and producer Steve Fleischman had woven our stories together with the hope of helping to trigger some environmental concern. Back in 1969 some nation-wide interest was badly needed.

One of the first places we visited was Central Park, the grand-daddy of all American city parks. I found myself wondering what New York would be like today without the park, and being grateful to visionaries like Frederick Law Olmstead, its early planner, and others like him who launched the practice that had helped preserve some trees in the midst of our cities' concrete canyons. The pam-pered dogs from penthouses who are gathered up in fistfuls by professional dog walkers each morning must also breathe a sigh of relief.

Later we ended up at an aquarium at Coney Island. The beluga whales were making playful faces through the glass in their tank at Mary Margaret while she posed for pictures in front of them. I slipped away from all the photographic commotion and went topside to the out-door seal tank. I came up quietly behind a group of teenage boys who were gathered at the edge of the tank, taunting a huge sea lion that had climbed to the top of a rock in the enclosure. The animal was completely oblivious to the kids below, who hurled insults and objects at him, trying to provoke some response. He sat immobile, perched on the rock, with his head and neck stretched to the side as he stared over the board fence toward the ocean beyond. One of the kids found a good-sized rock in the weeds near the fence, exclaimed loudly to his friends and picked it up. This ought to make the critter move. He turned toward the tank and suddenly saw me watching him, camera in hand. He grinned self-consciously, tossed the rock back in the litter-strewn weeds and walked off quickly with his friends.

Coney Island. Could this really be the beach Walt Whitman used to wander along, searching for inspiration and refreshment in leaves of grass? As if in answer to that question, I saw a metal sign in the tiny weed and rock patch that was located between the asphalt deck and the board fence that shut out the view of the sea. The boy's rock had landed nearby. Old, rusted and tipped to one side, the sign had been placed by another visionary with a sense of irony:
"LONG ISLAND ONCE USED TO LOOK LIKE THIS."

BEULAH BAPTIST BOOGALOO

"God Bless You, Brothers and Sisters, and AMEN!"
The black preacher finished the emotional sermon, collapsed into

the cape held by a deacon, turned and left the pulpit, helped away by the attendant.

Suddenly, he threw off the cape, leaped back to the pulpit and hammered home another thought. Another chorus of "amens" and he once again put on the cape. The choir jumped in and filled the fragile walls of the little Beulah Baptist Church of Eggbornsville, Virginia with sounds of love and hallelujahs.

Shades of James Brown! I turned to Chuck Perdue, a folklorist now at the University of Virginia, who had invited me to attend this Homecoming service in the backwoods of Virginia. Chuck explained that soul singer James Brown uses a cape in his concerts, probably learned when he sang in a similar church choir.

Homecoming was a perfect time for rejoicing and touching roots, even if they were just musical for me. Blacks had returned from all over the nation to this small home church of their youth; a visiting choir from the Peace Baptist Church of Washington, D. C. really added joy to the occasion. The singers flowed with the mood, improvising and rejoicing together, completely capturing the three white visitors who sat, tapped their feet and sang along in the rear of the church. Their music was fluid and free, drawn from rhythms and roots that brought them through periods of darkness and have so strongly influenced the musical traditions of America.

You dig the MoTown Sound? Check out Beulah Baptist Church sometime. You'll see where it came from.

"Something Big" had come home. Its rhythm and movement somehow reached through our great commonality back to small churches like Beulah Baptist. A part of me had come with it. The song had sprouted in San Francisco, but its roots were here.

SOMETHING BIG

You can't travel very fast in a Volkswagen camper, thank God; it's easier to see a lot more that way. However, one night I saw too much. It's haunted me ever since.

While driving north on Interstate 95 between Fredricksburg and Quantico, Virginia, sometime after midnight, I was fighting back sleep with some songs sung earlier in a small coffeehouse in Fredricksburg. The Interstate was lightly traveled; the blackness of night around the headlights' puddles whirled like a trance, chorused by the engine drone and transmission whine. I opened the window to let the night air shock some of the sleep away and checked the rear view mirror. A small set of headlights was growing rapidly closer in the lane to my left.

Suddenly, far ahead on the fringes of my own lights a beautiful white-tailed doe floated over a guard rail and lit in the middle of the left lane. Her hoofs slipped awkwardly on the road, spreading her legs far apart. Head down, the terrified doe sniffed at the strange substance that intruded between her feet and the earth, imprisoning her. She was blinded by our headlights, couldn't get her footing. The other lights raced closer in my mirror. I hit my flashers, turned out my lights and slid to a stop on the right shoulder, giving the onrushing car my lane. If only he could see . . .

"My God! He's goin' a hundred miles an hour! The sonafabitch!" I yelled and honked the horn, trying to jolt the deer into action. The station wagon flashed by, roaring upon the petrified animal still pathetically sniffing between her spread forelegs.

The car's brake lights stuttered. There was no chance for him to slow down, let alone stop. The doe was swallowed from my view, the whole terrible scene suspended in time—a split second. Obviously panicked, the station wagon's driver stomped on the accelerator and raced ahead, carrying the lifeless deer pinioned on the front of the car.

"You dirty bastard! Stop!" My headlights were on bright as I roared down the road. They were the only part of the Volks that could touch the fleeing car. I cursed the slow contraption that propelled me on a hopeless mission of revenge. Then, at the farthest edge of my lights, I saw the crushed deer slide off the fleeing car, spin across the right lane onto the shoulder like a rag doll. I abandoned the chase, pulled in behind the body, trailing the bloody smear that marked her passage, and wept.

My rage slowly passed and I quit hating the unknown driver. From the frieze in my memory I could recall what had appeared to be a family in the car. Probably hurrying to D.C. after a weekend at the beach. A weary, bleary-eyed father wanting to get some sleep before battling rush-hour traffic the next morning.

The incident with the deer gave me many of the same feelings and apprehensions I get when confronted with a large American city. Our cities seem to be especially saturated with the consumerism and technological jumble that is leading us all headlong into a future we cannot or dare not contemplate. Like the speeding occupants in that station wagon, we fail to make connections with what lies ahead, and hurtle through the darkness hoping to survive the ride. We know instinctively that much of what we are about is wrong, but the hammering on our nervous systems is much too hard to bear. What can the average person do anyway?

The question is, can we as a people learn to accept the consequences, or better still forestall them, adapting our living standards and applying our ingenuity and industriousness toward solving the problems that confront us in the preserving and restoring of our land, can we combat the racism and repression that divide us? Happily, many indications I've seen in recent years say yes, the people are ready to try, especially the young people and, paradoxically, people in the cities who have had to deal with our errors longer than those sheltered in the countryside.

There is vitality in the city waiting to be tapped. "Something Big" is drawn from the optimism I've shared with so many city folks I've met who are already about the business at hand. These people provide the roots for the hope expressed in "Something Big." There have always been those who have known what we are about.

LEROY

Leroy probably never heard of General Davis, probably never will. He did learn to orienteer, however, to break out of Hunter's Point for a day and run free through eucalyptus groves and along clear streams in the mountains above San Francisco.

With Leroy and his friends I supported the dream of General Raymond G. Davis, the Assistant Commandant of the Marine Corps. He wanted to introduce the sport of orienteering to the United States as part of the Marines' contribution to the Bicentennial.

Orienteering combines contour-map and compass reading with cross-country running. It's a healthful, inexpensive sport that helps introduce people to the outdoors and builds confidence in their ability to navigate on land in any season. Each person sets his own pace and chooses the course length he wants to run. Runners locate checkpoints on the ground that are marked on their maps.

The dream caught the imagination of many young officers and enlisted men in the Marine Corps. It was a stateside community activity Marines could share, particularly with underprivileged kids in areas near Marine units. It would sharpen Marine skills and act as a positive bridge to civilians. I was optimistic because I fully expected that General Davis would be appointed the next Commandant and carry the dream through. A welcome visionary.

In 1972 the Marine orienteering team worked with Universal International to produce a film about orienteering called *By Map and Compass*. It showed two groups of kids from the Mission and Hunter's Point areas of San Francisco learning how to orienteer.

Leroy was a member of the Hunter's Point Boys Club. I remember him best because he was the shortest guy on the team. What he lacked in size, he made up for in determination.

"Run, Leeeeroy, run!"

The earlier finishers were yelling encouragement to a small knot of runners that rounded a corner in the old dump we were using as a training ground. As the kids ran toward the camera filming them near the finish line, Leroy's short legs pumped wildly, trying to keep pace. He fell slowly behind, but didn't quit. A huge puddle loomed in front of the camera. The other runners detoured around it. Not old Leroy, though. He saw his chance and splattered over the puddle like a coot on take-off, and beat the others to the finish. His friends went crazy.

"Hey, man, did you see that dude run!"

"Hoooray, Leroy!"

The brothers from Hunter's Point gathered around their panting, mud-speckled teammate, laughing, congratulating him. They were a tight group.

Hunter's Point felt like a concentration camp to me. Little or no green space, it is comprised of ramshackle, former military housing units and walled in by a naval installation. Many of these kids had never been outside of the city, let alone in the wilds. They were even isolated and confined from the rest of the city. Shut in. Hardly surprisingly, there were more Panther posters on billboards than established political placards. Joseph Alioto (then Mayor of San Francisco) was framed by AK-47's.

Mas Vida, San Francisco

I remember watching a strange dog being attacked by local dogs of the neighborhood down the street from the boy's club. He'd wandered into the wrong territory. The neighbor dogs quickly hamstrung him and kept pressing the attack even though the intruder had tried to submit. A couple of boys curiously watched the proceedings much to the delight of their attacking dogs. I watched the affair from the viewpoint of another outsider up the street and wondered why the kids didn't stop the slaughter. Instead, the prey and his attackers thrashed out of sight around a building taking the spectators with them. I don't know if they finished killing him there or not; wasn't in a position to see, let alone understand.

The kids of Hunter's Point and Mas Vida quickly learned the rudiments of orienteering. After they were trained in alleys, junkyards and dumps, it came time to take them to the mountains for the meet.

"Hey man, when they gonna lay some food on us?" A young brother wearing a black overcoat and grey fedora jived up to me, accompanied by a large transistor radio he'd brought to the forest.

I pointed out where the lunches were kept and watched him saunter in that direction. Most of the kids, however, had stayed fairly bunched up and hadn't wandered very far from each other. After they'd finished eating and had a few moments to themselves, some began to regard their new surroundings with interest. One young fellow picked up a

Orienteering training ground at Hunter's Point, and later, in the forest

Running with map and compass in the stillness of the hills

club-like bone from the skeleton of a dead cow and seemed lost within himself, a blissful Samson.

While they waited for the runs to start, I wandered around with my camera, completely captivated by these kids. They had a character and integrity that was almost pensive. They were aware—real. Simpatia. Learning early they were faced with a deck stacked against them, they had found solidarity and happiness in situations where someone like myself might have despaired.

Suddenly, they were on their own, away from the clusters of their friends. Running with map and compass in the stillness of the hills, with only the sounds of their own breathing and footsteps. Through a telephoto lens I watched the tiny figure of Leroy bobbing across a green field under a massive sky. Alone with himself, "running on a rainbow."

One by one they all left the starting gate, pursuing the elusive checkpoints that were recorded on their mysterious maps, as strange new sights and sounds unfolded before them. Soon a rhythm developed in their running and each one settled into his own pace; a pace that allowed him to see more and feel a part of what he was running through.

Many weren't really ready, though. They had been trained too fast and hadn't had a chance to get in shape before filming. One by one they straggled through the finish gate, proud but exhausted. They were suddenly a lot quieter than when they'd first climbed off the bus. Were they just tired, or had they run across parts of themselves they'd never met before on those quiet trails?

It was getting close to evening. Quietly, and even a little reluctantly, many turned in their compasses and the numbered vests they'd worn that day. Then we climbed on the bus and headed back to the city.

As the bus droned its way home, I found myself wondering if these kids would ever get to run in the hills again, or if this would be a one-shot experience for them. The kids were being used in a way; there was a strong possibility that they would be forgotten after the film was completed. These fears were quieted by General Davis's dream, however, because Marine units in the area would keep helping the kids get a head start in pioneering a great national pastime. A world sport that would eventually be included in the Olympic Games.

Headlights flashed through the bus windows, skimming over the tired kids who were sleeping, quietly talking or thinking; I was watching them in the dark when I noticed one of the older kids holding something. It was a compass he hadn't turned in yet. As we rode along, he stared at the compass, turning its dial while the soft luminescent numbers glowed in the dark. He looked up suddenly and noticed me watching him. A little flustered, he started to hand it across the aisle to me. I shook my head, motioning him to put it back in his pocket. Even if the Marines didn't come back, at least there would be one compass around Hunter's Point to help somebody find his way.

It's probably a good thing, too. When I got back to D. C. in February of 1972, I was stunned by the news that President Nixon had not recommended General Davis for Commandant. Rather, he selected one of his former aides, General Robert E. Cushman, Jr., who as Deputy Director of the CIA had supplied burglary paraphernalia to the White House plumbers for use in the Ellsberg break-in.

Davis would retire. With his passing.would go the ideas for actively promoting orienteering for the Bicentennial.
When I queried Marine Corps headquarters two years later, I found out the dream didn't survive the change of command:

"There are no plans to promote orienteering in any special manner for the Marine 200th Birthday or the Bicentennial. Rather, there will be a 30-minute official film about the history of the Corps for late spring '75; a standardized 20-minute audio-visual presentation for early spring '75; a series of TV and radio spots and posters; MarCorp publication of books on Marines of the revolution; 14 historical paintings (with prints available for purchase); a joint DoD exhibit van touring parts of the country; and a joint-service Bicentennial Band."

Happy Birthday, Leroy.

Something Big

Cracked sidewalks and red fire hydrants,
Streetcar trolley rolling through the square,
Pigeons a-flying, newsboy a-crying
Something big is moving, moving through the air!

Morning comes to the city in a flurry,
People begin again to begin.
Chasing dreams
Planning schemes,
For this could be that special day,
A day to look and start to see,
And move up towards that something,
Something moving big!

Start now to seek the silence of your spaces,
Deep within yourself a symphony
Faces passing,
Inside you're asking,
What is this that's calling me,
Down boulevards and through the streets,
Riding elevators up,
Climb, climb, climb into the air!

Leroy

Can you see that I am running on a rainbow?
Would you like to rise and go with me?
Throw away your fear,
The time is drawing near.
Won't you try to stop and see,
That I am you and you are me?
Through love together we can touch,
Touch and catch that something moving big!

10. Gypsy
of My Mind

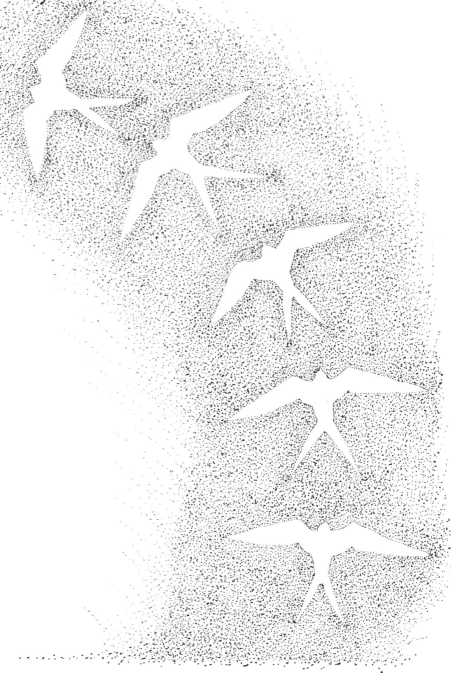

OKINAWA—SWEET POTATO

I can't remember her name, but she was beautiful. Carved by a life of hard work in the fields of Okinawa, her hands were gnarled and her shoulders stooped. Her lively, incisive eyes were warmly set in wrinkled, sun-darkened features.

The pineapple she sold me along the roadside smelled good, but I didn't have a knife. She understood and waved for me to follow as she trotted down the tree-arbored path to her hut. Sliding back the door, she slipped off her sandals and motioned for me to enter. I left my shoes at the door, stooped low and sat cross-legged on the tatami mats while she put on tea.

She hurried to peel and dice the pineapple in the soft light of the shoji screens. An open panel revealed the transparent green of vines and bushes tossing gently in a light sea breeze outside. The stillness was everywhere, broken only by the musical bubbling of the tea kettle. Soon all was ready.

A hot cup of tea, pineapple chunks bristling with toothpicks, a precious plastic bag of sugar. The old woman smiled across the table. We communicated through our eyes and hand gestures. I groped for some kind of word that might convey how I felt at that moment.

"Utsukushi, utsukushi." I offered the Japanese word for beautiful. She nodded eagerly. After we finished the pineapple, she continued to watch me curiously, knowingly. I relaxed and let the tea work its magic as the broth of humanity flowed between us. Binding us.

WHAM! KAAROOOOOM!

Everything shattered momentarily as a flight of Phantom jets screamed seaward toward China, rending the air. I was angry at the way a timeless, perfect moment had been interrupted. I guess it showed. She probably noticed. It was easy to calm down, though. The transient, outmoded silver beasts were really just a quick flash in time's pan. Soon they would be gone; the old hands

The old lady of Okinawa

that held the cup across from me would always be here. Her smile was eternal. Everything settled back into place.

When it came time for me to go, she left the room and brought back two plump sweet potatoes. At first, I dumbly thought she wanted to sell them, but she refused money. They were a gift. I didn't want to impose further, so I respectfully declined. Again, the knowing smile. She replaced the potatoes and we said goodbye.

The significance of her gesture didn't become apparent until months later at Obon, the celebration honoring the return of the ancestral spirits. During Obon, families unite for feasting, dancing and singing. The departed loved ones' spirits are considered benign and the occasion is a joyous one, often marked with the exchanging of gifts. Street dances accompanied by flutes, drums, samisens and gongs often last all night.

Like all other festivals, Obon was celebrated that year by large numbers of young people, not just the older ones. Together they helped the gentle flame of the Ryukyuan soul continue to endure and spread beneath the surface veneer of foreign occupation, as it has throughout time. The spirit of the land and its people continued to whistle like invisible bull-roarers of the unconscious, haunting those who descend on these islands. The Okinawans learned long ago not to stand against the typhoons of destruction and ambition that have swept

Obon dancers, Okinawa

back and forth over their homeland. Rather, they live beneath them, unseen but surely felt, until the typhoon moves on its way. The faces of the young people carried the timeless spirit in their celebrations. A spirit fed by the land, the springs, the sea and all nature. The mother they venerated. Earth. It was a spirit that could never die.

I often saw the humble sweet potato enshrined in a place of honor during Obon, offered with other foods to the spirits of the dead. I learned this is the food of life for Okinawa, having sustained them during famines and typhoons. In Naha, the capital, a shrine had even been erected in honor of the sweet potato and the man who introduced it from China in 1606, Noguni Sokan.

Although I hadn't fully understood my friend's intentions on that quiet day of tea and pineapple, I've often thought about her, holding eternity in her hands and offering to share it with me.

GYPSY OF MY MIND

Okinawa was a staging base for American military operations in the Far East during the Vietnam War, and although the Marines were withdrawing from Vietnam at the time I was there, the effects of this national tragedy were still rampant on Okinawa. Drug problems, race riots, murders, and sagging morale in all branches of the Service were aggravated by a growing restlessness among the islanders, many of whom resented, yet were economically dependent on, the American presence. The agony of the war and the national introspection we were beginning to experience were particularly apparent among many young Americans in Okinawa. They wanted to know what we were doing and why. They found their own style of living thrust up against a more ancient and timeless tradition and, like many Ryukyuans on Okinawa, experienced upwellings of feeling that what we were about needed improvement. Some sought relief in the ephemeral honky-tonks and whorehouses clustered around the bases; others turned to the solace of the eternal countryside to seek their meanings. One such place was a tiny offshore island called Hamahiga that attracted many of our young Marines. It didn't take long to find out why so many were drawn there.

Acting on the suggestion of one of my reporters in the Public Affairs Office, I hopped a boat at Yakena and joined island commuters returning home from Okinawa. About two miles ahead lay Hamahiga, flanked on the north by her twin sister, Henza Island.

Unlike Hamahiga, which at that time could only be reached by boat, Henza was linked to the main island by a causeway constructed from blasted coral, built by Gulf Oil interests that were locating on Okinawa. Extending east from Henza into deep water was the umbilical cord that allowed supertankers to disgorge cargoes into enormous oil storage tanks that were sprouting all over Henza.

Throughout Okinawa one could see the emblems of Gulf Oil and Caltex. A refinery was being built by Toyo Sekiyu (Eastern Oil, a joint venture of Caltex and Ryukyu Sekiyu) to help fuel the industries of Japan. Many Okinawans feared that once the United States re-turned the island to Japan, many polluting Japanese industries would be sent here, threatening to befoul her lovely coastal reefs and waters.

Already Kin Bay, west of Henza, was polluted by a combination of factors, forcing closure of Ishikawa Beach, a military recreation area, to swimming. Poisonous sea snakes were plentiful and increasing because they are air-breathers; gilled marine life and the reefs appeared to be declining. Some mysterious imbalance had unleashed crown-of-thorn starfish that were killing the coral reefs—reefs that protect the island from the relentless pounding of the Pacific.

There was an industrial typhoon hovering around tiny Hamahiga. Japan and Taiwan were disputing ownership of the Senkaku Islands that lay 260 miles west of Okinawa and 100 miles northeast of Taiwan, where major offshore oil deposits might be located. Red China added her own ominous rumblings. She had also decided to enter the offshore drilling game along her coast and was buying up equip-ment.

CONOCO, AMOCO, Gulf, Oceanic and Clinton were American oil companies that Taiwan would enlist to search for oil around Taiwan in the future, and by July 28, 1970, Taiwan had contracted Gulf Oil to explore for oil near the Senkakus.

Antagonisms really began to flare when the United States an-nounced it was including the Senkakus with the Ryukyu Islands for reversion to Japan. Taiwan was outraged; Red China fumed; Japan grinned. The United States found itself caught between the con-flicting ownership claims of two allies. Yet, there might be oil in them there ills, so where to place it?

In a diplomatic move taken right from a high-wire act, on May 15, 1972, the United States adroitly slipped the Senkakus administra-tively to Japan, subsequently made a unilateral announcement stating it took no position on the different territorial claims of the contestants, and eventually warned American companies they entered the waters at their own risk.

The actions of the oil companies were even more adroit, how-ever. Gulf's open-field running was a good example. While helping Taiwan explore for oil near the Senkakus and elsewhere, Gulf had located earlier on Okinawa, before reversion, to elude Japan's national investment laws. While the American military umbrella on Okinawa watched the approaching typhoon, Gulf was shining up its big storage tanks next door to unexploited Hamahiga.

I looked over at Henza Island as the boat drew closer to my destination, and wondered how much longer little Hamahiga could slumber so close to the eye of the typhoon.

Stepping ashore on Hamahiga, however, was like returning to the past and a gentler, saner expression of life. The pace was slow, flowing. As I walked along the tree-lined, white sand paths through the little villages, I passed tranquil gardens and homes nestled behind coral walls. The people were open and warm. I was already beginning to feel that sense of peace that is missing in military life.

No cars glutted the narrow winding paths. I saw only one small Datsun pickup that was used to haul supplies around from the dock. Foot trails connected the villages. There hadn't even been telephone communication between the three small villages on the island until a bunch of enlisted Marines "acquired" enough battery-powered army field phones, poles and com-wire to connect the houses of each village mayor. I soon found that this kind of action was commonplace on the island. For over ten years, various enlisted Marines from Okinawa had been living quietly and working with the people of Hamahiga in their free time. All over the island I noticed gentle reminders of these who had passed through.

I cut across the island on a trail and passed the elementary school. From inside came the music of a Chopin sonata. I stopped to look in and saw a pretty schoolgirl sitting at a new piano, completely absorbed, lost in her music. The school's principal came up behind me to say hello and alerted the girl to her unnoticed audience. She stopped playing and smiled self-consciously. I encouraged her to go ahead. She grinned playfully and started banging out "From the Halls of Montezuma." The principal and I cracked up.

Then he told me the piano's history. A young Marine who had worked on Hamahiga years before had grown fond of the people. After he rotated back to the States, he wrote to the principal, and sent money in each of his letters. The principal banked the money and one day used it to buy the piano.

I looked around in time to see a group of Marines carrying a huge cardboard box toward the school ground.

They opened the box for the principal. Inside was a new power mower to use on the schoolyard the Marines had laid out earlier.

"Where did you get the mower?"

One of the Marines smiled and gave me some story about the un-

Lili in Okinawa

responsiveness of channels, etc. Finally a friend of his had apparently grown exasperated and had purchased the mower personally before he left for the States. His friend had asked him to deliver it for him.

The principal was very excited. He went into a long oration about all the fine things the Marines had done on the island. He ended with a hint about how nice it would be to have some more field phones to connect the two island schools. The Marines grinned.

"Incoming."

I continued toward Hama. Once there, I passed the junior high school that had been partly built by the Marines. In a corner of the schoolyard stood a monument honoring a deceased Marine officer, Colonel K. H. Shelly, who had initiated the Hamahiga project.

A piano, a lawn mower, a school. From those who wanted to leave something of themselves behind, men like the young Americans now sprinkled over the island, draped with clusters of kids that tagged them everywhere. They were all refugees, in a way escaping from the frenzy of the towns that crouched around the bases. The brothels, bars, pawnshops and steam baths of Okinawa seemed light-years away. We were momentary castaways enjoying being human with folks just like those back home, on a sand and coral speck of sanity that bobbed on the edge of a whirlpool of geopolitics.

On a calm, still pool
of unfathomed depth
floats
a small, green leaf
—her soul.

The understanding of the old woman, her willingness to share the spirit of her life with me, and the camaraderie of the Marines and Okinawans notwithstanding, my travels had been cast against a backdrop of loneliness and occasional homesickness. I'd left behind me a young wife, a woman I'd never really had the chance to know. The sharing with the old woman was an echo of a sharing I was missing. Lili's visit to Okinawa helped change all that. We explored the islands on a second-hand motorcycle, and closed up distances together. "Gypsy of My Mind" was written to welcome Lili when she flew to Okinawa to share my wanderings.

Gypsy of My Mind

Gypsy, Gypsy of My Mind,
Now we sit in different times,
Watching walls of sea and space,
Tumble down like windfalls of our pride.

But my mind's been changing with the times,
And I see you now with clearer eyes,
While I thank God that you're still here,
To give me hope and stay to share my life.

It's been a long time comin', baby,
But I guess it's better this way.
It's been a long time comin',
But I know our love is growin',
Gettin' richer every day, richer every day.

The coral reefs and cycle rides,
Without you here to share my life,
Joined lonely nights to probe my soul
And make me ache to see and touch your eyes.

Hey little girl with gypsy eyes,
Won't you step inside my mind,
So I might learn to know you more
And love you always, Gypsy of My Mind.

11. Windsinger

While many young Marines sought their meanings on Hamahiga, the games of geopolitics and war kept their leaders occupied—though not necessarily out of trouble. An earlier imbroglio came to mind while I was sitting at a hilltop cave-shrine near Kaneko village, on Hamahiga. The incident seemed even more absurd on that quiet afternoon as I mentally juxtaposed it with the efforts of the young Marines I watched below.

Major General Louis H. Wilson, commanding general, Third Marine Division, had been the main player in the drama of 1970. The same General Wilson who would be appointed Commandant of the entire Marine Corps in 1975 through the inscrutable wisdom of President Ford and the United States Senate, thereby succeeding previously-mentioned General Cushman of Nixon fame.

While commanded by Major General Wilson, the Okinawa-based Third Marine Division carelessly built a relic of the Vietnam War called a fire support base on leased lands in the mountains of the Northern Training Area. An Okinawa hilltop had been scalped of vegetation and 105mm howitzers emplaced in gun pits. Coils of concertina wire surrounded the emplacements and bunkers. Just like in 'Nam. The whole activity had been done rather quietly and hadn't attracted too much attention because of the remoteness of the site. However, when we announced to the press that the Marines would "have a round out of the tube" before New Year's, all hell broke loose.

The mayor of Kunigami-son, Takeo Yamakawa, complained that his people lived within range of the guns. They might be hurt. Live shells might linger in the forest soil.

Item: (He was right. There were civilians in range.) Guns would fire high trajectories. Safety precautions would be taken. Besides, we leased the land. He couldn't tell us what to do on it.

The mayor soon received support from the Japanese media and various island unions and organizations. Queries began to flood the public affairs office.

"What about forest fires? The steep mountains are thickly wooded and are good watershed."

Item: (They were right.) The Division said they would fire variable time fuses that would detonate the shells above ground. (That is, if the fuses worked. VT often has a strange habit of malfunctioning; it then acts like a regular artillery shell. Having fought many fires in the States, I sure didn't envy future Smokey Bears in that terrain.)

"Don't you have enough lands for practice already? You already have a big firing range. Why do you need more?"

Item: (They were right. We had a large range and didn't really need this one.) But Major General Wilson wanted to have a round out of the tube by New Year's.

The Okinawan coast

"What about the woodpecker Noguchigera?"

"What woodpecker? Who's Noguchigera?"

"Noguchigera is the rare Okinawan woodpecker found only here. He's on the endangered species list. Didn't you know you dumb Marines put your firing base smack in the middle of an official wildlife sanctuary?"

"Ohmygawd! We better tell the head-shed!"

"What woodpecker? I don't see no woodpecker."

The Okinawan press scurried through the hills and quickly got a picture of one. Bird lovers around the world rallied. The confrontation grew.

As D-day neared, our friendly contacts in the Okinawan press kept the public affairs staff informed about what was developing. We informed Division. They acted suspicious. How did we know things that Division G-2 (Intelligence) didn't? Personnel from our office even attended an Okinawan mass meeting and watched their battle plans being laid. The Okinawans were tired of being shoved around. Noguchigera had become a battle cry—an environmental crusade.

When we informed the Division that 600 people would be coming through their barbed wire if they didn't back down, they wouldn't believe us.

There would also be thirty-five or forty accredited newsmen from all over Japan and Okinawa to record the event. Division leadership told the public affairs staff to stay off the hill.

However, pretty soon the cement-heads in Division began to see the handwriting on the wall: "Tell 'em we're only gonna fire blanks."

Item: The guns will only fire blanks on New Year's Eve.

"No dice, Marines. The noise will frighten Noguchigera. His nesting grounds will be disturbed."

Item: Major General Wilson was adamant. The Third Marine Division *will* fire blanks on New Year's Eve. (Hmmmmm. What good are blanks to an artilleryman? I thought they were only for parade grounds . . .)

Things had quickly deteriorated into a blind determination to follow a bad decision through to the bitter end.

We sat exiled in the public affairs office on the morning of New Year's Eve and waited for the hotline to start ringing. (Bearers-of-bad-news slurping coffee and holding their heads.)

"Situation report from the Northern Training Area: There are about 600 Okinawans surrounding us . . ."

Pass the sugar please.

"Situation report from NTA: The 'trespassers' have us surrounded. A demonstrator is flying the Rising Sun flag from a tree by our wire . . ."

Do you take cream?

"Adverse incident report from the NTA: The Marines ordered the demonstrator to come out of the forty-foot tree. He refused. We told him we would cut it down. He still refused. We cut it down, but it fell down the hill. The injured demonstrator has been medevaced to Kue Hospital."

Those meatheads! The demonstrator was in critical condition: punctured lung, broken leg, lacerations. A momentary standoff had been turned into an ugly rout. The Okinawans, backed by a huge corps of newsmen, stormed the wire and attacked a small contingent of unprepared Marines inside the fire support base. The Division flew in helicopters to snatch the howitzers from the demonstrators' reach.

The Okinawan media were on the phone. They wanted to know what was happening. It would take all day for their own reporters to get back to Naha and it took hours to set Japanese type. We prepared a release stating what had happened. The Division ordered us not to say anything. (Not to say anything when the whole world had been there.)

We stuck our necks out and released the story anyway, asking our Okinawan editor friends not to roll the presses until things calmed down at Division. They would hold the presses. Late that evening Wilson gave permission to release the story that was already set in type for the morning editions.

When the story broke we found the Okinawans had used some of our information. In fact, I got the feeling they had been pretty kind to us. However, there was one photograph in the paper that told it all. In the foreground a group of Okinawan farmers in coolie hats carrying the Rising Sun flag framed a gun pit. Over the sandbags hovered a Marine helicopter carrying away one of the 105s. It was time to "study the situation."

Item: "Major General Louis H. Wilson handed a letter to Mayor Takeo Yamakawa of Kunigami-son which read in part, 'I can now inform you that the Marine Corps has revised its plans and will utilize this site to familiarize troops with the characteristics of a fire support base, without live firing.' "

It was over.

Noguchigera had won. The Third Marine Division and future Commandant of the United States Marines had been defeated in open combat by a woodpecker.

A family of Okinawans was climbing the stairs to the cave-shrine near Kaneko. They carried food and offerings for the spirits that inhabited the place; spirits I'd been sitting and thinking with. It was time for me to leave Hamahiga.

Hamahiga Island

It was dark when the boat pulled into Yakena. I climbed on my motorcycle and rode along a sea wall north of Gushikawa. Soon I stopped the cycle, turned off the motor and looked across the dark bay. Henza island was aflame with floodlights and activity while Hamahiga slumbered quietly in the darkness at her side. What happened to Henza, Hamahiga and Okinawa prefigured the world's future. The same forces of avarice and apathy that had destroyed the American Indian were at work on the gentle Ryukyuans.

Like the Navajos, the Ryukyuans were victims of an out-of-control population explosion that rapidly closed off their options and forced them toward industrializing their lands. Like the American Indian, the Ryukyuans had been pawns in the storms of ambition and conquest of others. And like the Japanese, they were being seduced by industrial consumerism and materialism that would transform their culture. They would soon be part of the economic "Great East Asia Co-Prosperity Sphere," with Japan on top and the American military umbrella overhead. Gulf and Toyo-Sekiyu oil facilities were arriving, creating giant landfills. Gulf's causeway would finally kill Kin Bay. A four-lane freeway would be engineered by the Japanese to handle the many thousands of new cars registered each year.

Population growth would continue to rise, demanding more land. More people would need more jobs. Many would be forced to find work in Japan, yet many Okinawan workers would make a U-turn and come back to Okinawa, disillusioned with the values and frenzied pace of modern Japan. At home they would be torn and frustrated, overskilled and out-of-work.

The U-turn Okinawans reminded me of Navajo strip miners and highly trained drag-line operators at Black Mesa in Arizona. Sometimes these workers just fail to show up one day and, according to a mine official, "go back and herd sheep," leaving the whites, who invested great sums of money in their training, smoldering behind them. Apparently, a need even a good job can't satisfy exists there, too. Yet, ironically, I would meet Nanao Sakaki, a Japanese poet, at a Zuni Indian Shalako Dance in New Mexico. Nanao would tell me that he had left his homeland because the Spirit of the East is dying there, and is more alive now in the American southwest among the pueblos and deserts.

The experience seemed doomed to be repeated endlessly. Years before I had spent some time performing folk music at the Intermountain Indian School in Brigham City, Utah. "Windsinger" was drawn from experiences I had then. In the song the boy Windsinger sees the avarice and apathy that had crippled his own people conducted on a global scale. Spawned years before at a Bureau of Indian Affairs school in Utah, the meaning of the song inescapably surrounded me, hauntingly, on a small island in the western Pacific.

That night, wars of scarcity loomed in the future for the young Marines on Okinawa, unless we could summon the creativity to make technology a true servant—not our master—and blend the blessings of science and industry with our spiritual roots.

Could Okinawa endure? I knew that industry would demand more land. Agricultural traditions would be absorbed. The economic spiral would tighten, and soon hungry eyes would notice sleeping Hamahiga, unless . . . perhaps unless people hear the songs sung by a young Navajo boy approaching manhood, awareness. Unless they add their voices and work for a gentler, more balanced world.

The waves lapped gently at the sea wall as a breeze picked up. I felt a little chilled. Overhead, the stars had come alive. It would soon be the seventh day of the seventh month—Star Festival—when Althair and Vega approach each other. At that time, a Ryukyuan cowherd and a weaving girl, who have been separated for a year, would join each other once again, by crossing the heavens on a bridge of swallows.

Perhaps she would offer him sweet potatoes.

Windsinger

Windsinger, ride! Windsinger, ride! Windsinger ride!
Nit C' Hi Hatatih, Nit C' Hi Hatatih, ride!

A young Navajo came riding,
While at his back set the bleeding desert sun;
He sought his name on the mountain,
From the wise man called Naga-Khan.

Naga-Khan was such an old man,
And his eyes were filled with silence,
The silence passed from the ages,
When the spirits first walked and breathed upon the land.

The old man smiled at the boy,
And spoke with the strength of ten-thousand desert winds;
"Ride the Four Winds of the mountain;
Wake, and see, and think, and speak—WINDSINGER!"

He was carried to the North from the mountain,
To the land of the Bear, and silent, frozen faces;
The Bear sat and watched the Eagle,
Whose talons had been tangled in the darkness.

The Eagle tried to stretch his wings,
And to tear himself free from the darkness;
But his people were all sleeping,
And not one person cared to awake and cut him free.

The boy breathed deeply from the winds,
And sang with words that struck like burning spears!
"Hear your empty lives of deafness!
Awake now! Can't you see the danger?"

He was carried by the wind to the Southland,
Where the Lion and the Llama crawled on bloated bellies;
Their eyes were only blackened sockets,
And children tried to run while rickets swelled their knees.

The rich would feast and make their speeches;
Tomorrow would bring the needed changes;
But their speeches were all empty,
And the hungry children groaned 'til their parents could not sleep.

Then the groaning changed into a roar,
The roar began to thunder and the wind began to scream:
"Do not let the children suffer!
Make all the people wake and see—Windsinger!"

He was carried to the East across the ocean,
To a place where the Dragon spewed its venom on the land;
He saw the vacant stares of millions,
Hungry stares that held no hope for food in empty hands.

The Dragon crouched before the Eagle,
That had gorged itself on darkness,
That had slept itself to weakness,
In whose eyes grew dim the light that could still make all men free.

The spirits moved upon the Winds,
Till the voices of the ages began to echo in his ears:
"Ride again among your people;
Make them wake, and see, Windsinger!"

Windsinger rode among his people,
Singing in the cities, the suburbs, and the ghettoes;
He sang to his brothers in the hogans,
In the squalor of the reservations.

But the people would not listen;
And the children danced their frenzied dances,
As the neon whirled, and flashed, and blinded;
And poison-bearing clouds hung like incense in the air.

The boy sang out his broken heart,
Then with sorrow in his eyes he rode slowly to the West.
"Dance while you sleep, my people;
The songs that I sing have no meaning for your ears."

But as he rode he heard a rumbling,
For a few had woke and turned to raise their faces;
They came forth from the darkened ghettoes,
From the cities, and off the reservations.

A few had stopped to hear the singing,
And awakened from the slumber of the many;
Each one raised a burning spear,
With which to pierce the darkness and let forth the shining light.

The winds began to move and sing,
Till the valleys echoed with a hundred rolling songs;
The man rode to his sacred mountain,
To wake, and see, and think, and speak—Windsinger!

COMING HOME

12. Dick's Song

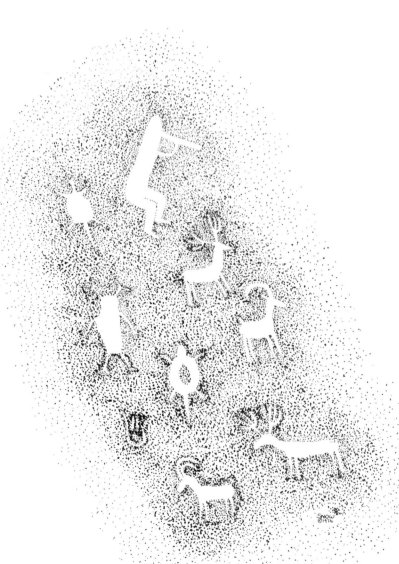

Dick Smith, bush pilot for Canyonlands Aviation, shouts above engine noise to his passengers, "Enjoy the view now, folks . . . won't be long and it'll be gone. Not like it is now anyway."

The passengers, with noses and cameras pressed against cabin windows, gape at panoramas of Rainbow Bridge and Lake Powell below. "Before the power plant at Farmington started up, we had over 170-mile visibility down here," Dick continues. "Now it's more like eighty. Just think what'll happen when five or six more plants like Farmington get fired up! You'll think you're right back in Los Angeles!"

Dick turns back to his controls, muttering to himself, "You may as well be too, since this power is for you and your damned air conditioners!"

"It's been a long time since you've seen the Old Girl, hasn't it?" Dick tipped the Cessna 180 up on a wing so I could once again stare in wonder at an old friend—the confluence of the Green and Colorado Rivers.

I'd just returned from duty with the Marines in the Far East and was hurrying across the country in 1971 when Dick hoodwinked me into pausing long enough to pick up some river runners at Hite, Utah. I was already behind schedule and would probably be late reporting to Marine Corps headquarters in Washington, but Dick had a way of making people forget about unimportant things. He didn't truck much with clocks.

We'd lifted off from Dick's little dirt strip at Canyonlands Resort, just east of the park's Needles entrance, and watched the colorful convolutions and spires of this mysterious country pass beneath us. The Needles, Maze, Doll's House, Land of Standing Rocks—old friends passed in review. They were friends made years earlier when Dick had helped me bridge the transition from the uplands to the desert. I snapped out of my reverie and half-seriously reminded Dick about my time schedule. "How long will this take anyway?" I shouted above the engine noise. "Duty calls, you know."

Dick answered with a grin through his big bushy beard, and shook his head mockingly from side to side like a furry pendulum. "I can see right now we've got to start your trainin' all over again." He cut power and pulled the plane's nose up. "Yesirree, we'd better go down and have a closer look at the Old Girl!" As the plane approached stalling speed, I saw Dick kick in the rudder pedal.

"Hold it," I yelled, "We're gonna cross-control sta-a-a-all!" The little Cessna flipped over on its back and we rolled nose down, spinning past the towering sandstone walls, plummeting toward the river. As millions of layered years rushed by the window, the sculptor of the

canyons, the Colorado—the great spiritual and physical backbone
of the American West—waited serenely to gather us in.

Not today! Dick pulled out of the dive just above the river surface and swooped up alongside the west wall. I watched our wing tip
race past the crumbled ledge.

"Rocks is fun!" he yelled. We were approaching another stall . . . a
hammerhead to the left!

"Yippeeee-shit!" I splurted as the plane broke and plummeted back
down the wall. We skimmed over the river and climbed up the east side.
Then another hammerhead, this time to the right!

"Terrific . . . " I groaned. As the river grew larger, Dick grinned at
his passenger. We leveled off at his former crop-dusting altitude and
cruised downriver, skimming above the water surface.

"What time did you say it was, anyway?"

"How the hell should I know!" I gasped, "I threw my damned
watch out the window on that last dive!"

"Good. Now you're ready to be welcomed home. Let's go get those
river runners."

Ahead the walls opened up at Spanish Bottom where the Spanish
Trail crosses a fordable stretch of the Colorado and twists up the canyon
walls to the Doll's House, sandstone sentinels that look down on the
last quiet stretch of river. Around the bend lurk the rapids of
Cataract Canyon, or at least the remaining fragments of wild rapids
that haven't been drowned by the backwaters of Lake Powell.

It was winter, so the roar of our engine didn't disturb any river
parties. The Colorado was resting from the previous summer's invasion
of rafters, wild-river enthusiasts whose numbers grow rapidly every
year. Dick was always drawn to the river, particularly the rapids.
Yet, in his pellmell existence and business pursuits, he had ironically
never gotten to ride down Cataract. A victim of his own clock, he
never found a clearing in his schedule to accept any of the continual
offers from his river friends. Or he'd end up giving up his ride to
impoverished freeloaders like me. My first trip through Cat was in
Dick's place with river guide Eli Gourdin of Tour West.

During the summer, river parties often looked up to see Dick
skimming above the rapids on a mercy mission to drop an important
message, or a cardboard tube full of ice cream, then wag out of sight
around a bend in the river. Another of his double-time
days . . . just checkin' folks. Howdy.

"Who's crazy enough to be runnin' the river this time of year,
Dick?"

"Who else? Old Tex McClatchy. I guess he's got the Aristotle
Onassis of Norway and his wife along. They wanted to ride the rapids

Below Big Drop, Cataract Canyon

even if it is winter, so they chartered Tex to take 'em. I thought we'd check the river for 'em on the way down.''

Ol' Tex. What a character. Dick, Tex and I had spent a lot of time terrorizing the countryside in years past. Who else but Tex would have guts and humor enough to try this? He was a wild-man river lover who shocked all of Canyonlands by building and launching a 100-foot paddle-wheel riverboat, The Canyon King, at Moab to supplement his offering of float and canoe trips down the Green and Colorado Rivers. Tex had found his jetboats too noisy and expensive to operate in the silt-laden, turbine-destroying waters of the Colorado, so he decided to step forward by going backwards and had adopted the quiet, nostalgic Canyon King, whose decks could handle 200 passengers at a time, at a fraction of the cost of his jetboats.

But eventually when the sandbars got too high, and the river too thick, Tex would wrench the 75-ton stern-wheeler from the river and drag it over desert highways more than 170 miles from Moab to Lake Powell where it now operates out of the marina at Hite. A new career for a stern-wheeler in the middle of the desert—only Tex could pull off something like that.

"Isn't that the mouth of Dark Canyon ahead?" I yelled over engine noise. "Hell, the damn lake's already covered up my old campsite!"

"Yep, she's gone under like all the rest." Dick put the plane in a steep climb and rose above the walls of Cataract so I could gaze up inviting Dark Canyon toward its steep source in the Blue Mountains near the Bear's Ears.

Some of my finest memories of the area were of camping along the Colorado on a white sand bar that protruded from the mouth of Dark Canyon. I spent Thanksgiving there once in the light of a full moon, savoring the night, watching silver ripples move and grow, eventually filling the entire river gorge with haunting, persistent light. My companions and I toasted the moment with gulps of hot tea brewed from the clear water of Dark Canyon, collected from the waterfalls and pools near its mouth. The next morning we spent gamboling and cavorting along the river, welcoming the warming sun rays that peeked their way down the chilly walls of the canyon. The sand bar was gone now, and we circled over the quiet lake waters that entomb it.

As Dick wheeled and dropped his plane back into the canyon and headed down-lake to Hite, we approached the start of the aqueous tomb of Glen Canyon, a place where exquisite beauty was sacrificed on the altar of the Bureau of Reclamation long before I'd turned from the mountains toward the desert. Like most Americans, I'd never heard of Glen Canyon and its marvelous wonders until it was gone. I envied those who had known her, like the families who had been able to navigate her in their own inexpensive rubber rafts. Unlike Cataract or the Grand Canyon, whose rapids demand the skill of professional outfitters or seasoned river runners, Glen Canyon was a friendly place where even beginners could leisurely float and explore sandstone canyons, grottos, and gardens for the over-150 miles between Hite and Lee's Ferry in Arizona. I'd first come to Lake Powell just as the dam's waters were filling some of these side canyons and grottos. It loomed below us now, undeniably one of the most beautiful bodies of water in the world, with a more exquisite beauty hauntingly beckoning below the water's surface.

Looking up, I saw the bridge at Hite spanning the canyon, cluttering Dick's flying room. He added a note of defiance and zipped the Cessna underneath the span and pulled its nose up to climb to the dirt strip on the cliff rim above. Tex and his passengers were waving from the strip, amazed to find a plane suddenly appear under them.

We landed and picked up Mr. and Mrs. Stolt-Nielsen of Oslo, pausing just long enough to howdy Tex, who was surprised to find me occupying the seat he'd expected to fly back in. After we roared off the runway, I settled back and prepared to describe the scenery to Mrs. Stolt-Nielsen as we returned to Canyonlands. There was one problem, however. We weren't heading back to Canyonlands. We were flying southwest, down the lake.

"Hey, ol' buddy, what's happenin'? I though we were just gonna pick up these folks so I could keep on goin'?"

"Guess I forgot to mention Grand Canyon. We're supposed to give 'em a ride down there." Dick grinned impishly. He didn't truck much with schedules either. He launched into his familiar travelogue, which combined with the awesome views to hold the Stolt-Nielsens spellbound. Nadia Stolt-Nielsen proved particularly sensitive and inquisitive about the desert. The river had worked its magic on her, and soon it would hold her fast.

I became more nostalgic as we flew, looking down on old memories as Dick constantly changed course, collecting points of interest on the long way to Grand Canyon. Happily resigned to the inevitable, I noticed Dick was grinning again. I wondered if the Stolt-Nielsens knew they were underwriting a Marine's homecoming tour. Suddenly we turned away from the lake and were flying up Escalante Canyon, a delicate wild canyon that shelters many places like those that were lost beneath Lake Powell. Dick was really orchestrating this flight.

"Remember the first time you flew up Escalante?" he called over his shoulder. How could I forget that flight of unbelievable events, centered around scattered thunderstorms with which we'd been playing tag in the

Colorado River, in the canyonlands

air. We had flitted over the canyons awestruck by the rainbows and cloud formations around us and the flash floods below. When we turned up Escalante it seemed like we were entering a forgotten jewel chamber. As we bounced between the thunderheads, light rays pierced the dark clouds and skittered across the water-filled potholes below. The surface of the misting slickrock reflected upward toward transient bits of life suspended in the fragile, buffeted aircraft. In the midst of the naked rock, the veiled, lush canyons looked like mysterious, primordial birthplaces.

We had been drawn like moths further into the heart of the storm, weaving through fickle gaps in the clouds that quickly, sometimes violently, shut behind us. Suddenly we were enveloped in fluctuating rainbows and iridescent clouds while the violence of the wind gusts below scooped and lifted the canyon waterfalls, tearing them into shredded veils of mist. The plane was like a brittle corpuscle in the center of a giant heart; the energy of the wind and storm reigned supreme here, and we were trespassers.

I loosened my seat belt foolheartedly to try to photograph the spectacle when the plane was slammed violently downward, thumping my skull against the cabin roof. It was time to get the hell out of there. We fled through a seam in the clouds, back down the canyon.

We hadn't retreated alone, however, A song followed us out, a persistent melody that grew into "Dick's Song," a celebration of storms, the river and the cycle of desert life.

But more than that, "Dick's Song" seems to be a frank admission of our own transience contrasted with the awesome and enduring reality of the rocks. Naked reality where the folds and anatomy of the earth stand revealed before us in a land of quiet scarcity that lets us escape distraction and focus in on things we may have rushed past before, like ourselves.

Awareness.

"Om Mani Padme Hum" is a Lamaist prayer that surfaced in the song's lyrics—the circulating Spirit of the East arising in the waters of a Western river.

Balance.

The prayer is a supplication for universality so that through the power and immortality of our own minds, we can probe deeper into our own reasons for being—like an unfolding lotus–center's growing beauty—and awaken to our roles in the universe.

Consciousness.

Dick's Song

Hey Dick! Hear the rain fall!
Smashing on the slickrock!
Cuttin' through the sandstone
Rollin' round the dry bones.

Roll you wild Colorado!
Rampage wild and free!
Carve your restless song in me . . .
I don't know why.
Time seems to pass me by,
Like the river that's rollin'
Through the silent shades of life.

Wait Dick! Hear the thunder!
Pounding like a heartbeat
Pumpin' down the flash flood,
Boilin' through the red mud.
The water carves the canyons
Through the technicolor years
And swallows up our questions,
Hopes and fears.
The river's prayin' "Om, Mani Padme Hum."
I wonder where it's rollin' me
And when the light will come.

Look Dick! See the sun now!
Look how it is streaming!
Piercin' through the dark clouds,
Shinin' on the wet rocks.
Rose-red, violet, burgundy
And bright orange comes the night
Across a land so vast and big,
It wanders out of sight.
Yet the sun still hangs around,

Till the night birds cry their sounds
Then resting on a mountain top
It lingers, then is gone.

Hush Dick! Feel the silence!
The muffled awesome quiet!
Walking like a dead man,
Movin' over dry sand.
The night is a velvet blanket
That wraps us in ourselves,
And makes us look to see our depths
Wherein the spirit dwells

Like the Anasazi etching,
His ghostly Indian paintings,
He tried to hold death's muffled feet
And keep his life from fading!

Hey Dick! Hear the rain fall!
Smashing on the slickrock!
Cuttin' through the sandstone
Rollin' round the dry bones.

Dick Smith

13. Hey, What's Happenin'?

Dick would smile a little self-consciously over his shoulder as we neared the song's birthplace. He loved music and would fly hours out of his way to listen to it. One morning, for example, some friends and I were staying at an isolated Utah ranch when we heard the roar of a plane overhead. When we looked out the window, we saw Dick's Cessna landing on the ranch's dirt road. Then Dick squeezed out of the cockpit holding his tape recorder. He ambled up in his Navajo mocassins, grinning, 285 pounds of airborne mountain man. Time to make music and record songs for the winter, when he would sit alone in his quiet canyons and play back memories.

Dick had a series of favorites that he always requested: "Suzanne," "Happy Wanderer," "Tyin' Knots in the Devil's Tail," "Buffalo Boy," "Color Crayon Morning," and others. He was always a little shy about directly requesting his own song, but his grin would break through his beard when we ended a session with "Dick's Song."

The weather was clear over Escalante that day. We had no way of knowing it would be the last time we would see it together. New kinds of storm clouds lurked out of sight. I'd first noticed them years before in 1968 when I visited Park Service Headquarters at Lake Powell as a graduate student.

I'd wandered into the engineering room during the lunch hour to examine some of their recreation facility maps for a research project I was doing. When I walked past an engineer's drawing table, I noticed a drawing in progress that bore a Park Service legend in the corner. It was a plan for a site called "Kaiparowits."

I stopped to ponder the situation, a little puzzled. It seemed strange at the time to see the Park Service involved in the Kaiparowits proposal to this extent long before the public had become widely aware of the plan, or before any research had been done to determine its environmental feasibility. I began to suspect that the power plant's construction was a foregone conclusion, and that the people would be the last to know. My suspicions were quickly confirmed:

"What in the hell are you doin' in here! Get away from that table!"

A red-faced Park Service employee angrily stormed into the room and hurriedly ripped the drawing off the table, rolled it up, and stuffed it in a metal cabinet.

No, he wouldn't discuss it.

Yes, it was none of my business. I grinned and turned back to the nonoffensive campground drawings. The engineer's disconcerted behavior had already told me more than I wanted to know.

Fortunately, the public would demand a closer look at the Kaiparowits proposal in years following, particularly after Secretary of the Interior Stewart Udall's "Bay of Pigs"—the Navajo Power

Plant—quietly appeared near the banks of Lake Powell, and began spewing its plume into the priceless airshed.

Kaiparowits promoters would receive a temporary setback in 1973 when Secretary of the Interior Rogers C. B. Morton held up the Kaiparowits project, realizing that the environmental impact of 3 thousand-megawatt Kaiparowits would combine with that of the 2.2 thousand-megawatt Navajo and degrade the quality of the area even further. (For example, Navajo reduced average visibility from ninety to seventy miles. A 1976 Environmental Impact Statement predicted Kaiparowits would lower it to fifty-five miles.)

But the power brokers were not deterred, and pressure to build the plant escalated. It wouldn't be until 1976, after the completion of an environmental impact statement, that pressure from the plant's opponents would combine with a decreased demand for electricity in southern California and rising construction costs to force Southern California Edison to withdraw its application, theoretically killing the project—at least for the time being.

"Dick's Song" would play a small part in Kaiparowits' *coup de grace* in 1976. Mary Belle Bloch of the Environmental Defense Fund asked me to join actor Robert Redford at Lake Powell for the filming of a television

Navajo power plant

documentary for the CBS television program *60 Minutes.* "Dick's Song" was used for background music, while the film presented some of the exquisite areas that would be endangered by the industrial madness symbolized by the proposal for Kaiparowits.

Kaiparowits proved to be just the tip of the iceberg, however. An industrial tempest unlocked by the stored waters of Lake Powell continues to descend on our beloved canyon country. Energy waste and unlimited growth in America contribute to a growing frenzy all along the upper Colorado River system that the Department of Interior predicts might eventually establish about thirty coal-fired power plants, coal gasification plants, and oil-shale developments dependent on great gulps of scarce water resources from the over-allocated and over burdened Colorado. (In 1973, the Bureau

Drowned grotto and side canyon in Glen Canyon, and Lake Powell

of Reclamation estimated energy developments would consume over 840,000 acre-feet of water per year by the year 2000.) Two of those facilities are planned or projected for the Escalante area, and would possibly draw their cooling water from a dam on the Escalante River, threatening the proposed Escalante Wilderness Area which is already threatened by a senseless road-building boondoggle that could slash its way across this priceless canyon jewel.

This is just a small part of the industrial cancer that will assault this great spiritual and recreational heartland of America. In all, the Department estimates up to nine coal-fired power and gas facilities are in progress, planned or projected to stretch from Glen Canyon Dam on the western side of the Green and Colorado Rivers northward to Huntington Canyon near Price, Utah, befouling the priceless national parks, monuments and recreation areas of Utah.

To the east of Lake Powell near Farmington, New Mexico, and the San Juan River another battery of seven coal-gasification units and three more coal-fired power units will join the infamous Four Corners Steam Plant that has spewn over 225,000 tons of pollutants annually. The *combined* effect all these plants will have on the Four Corners region air shed is staggering. Just the proposed Kaiparowits plant, that would have been located across Lake Powell from the existing Navajo Power Plant, would have dumped at least 320 tons of emissions per day (116,508 tons per year) in the Lake Powell air shed *if it had followed regulations and its equipment functioned*. The combined emissions of these plants will be concentrated by an ominous inversion layer that hangs over the Four Corners region at least nine months out of the year.

Prevailing winds from the south and southwest almost guarantee that fallout from the plants will land upstream from the dam in the river system. The fallout contains poisonous trace elements like mercury, lead and cadmium. Mercury is particularly deadly, and once on the ground upriver, can work its way back into the system and concentrate in the lake's ecosystem through bioamplification, as organisms higher up the food chain ingest the concentrated poisons in the body tissues of lower organisms. As early as 1973, before the Navajo plant was a factor, scientists had shown that, from natural run off, mercury levels in Lake Powell's larger bass and walleyes met or exceeded the upper safe consumption limit for humans (500 parts per billion), averaging 626 ppb, and warned that the combined effect of the Navajo and Kaiparowits plants alone could increase the fish levels as much as 150 ppb—not counting other upstream developments. The 1976 Kaiparowits Environmental Impact statement predicted that Kaiparowits alone would have added from one to twenty-seven per-

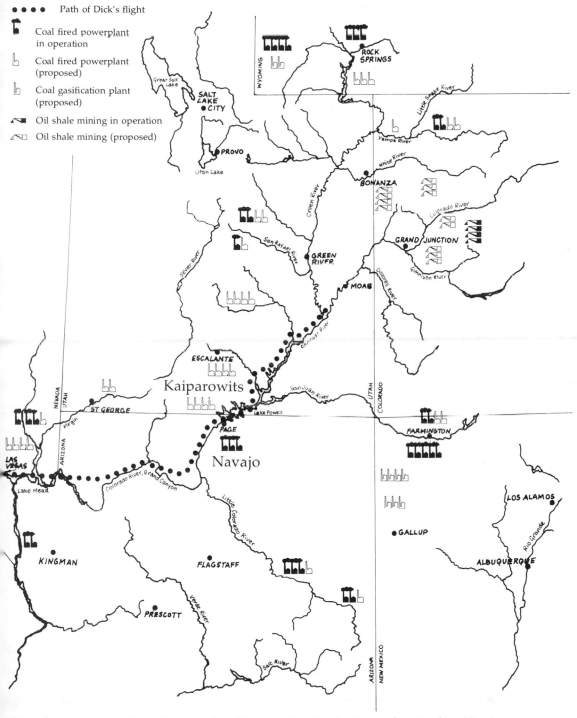

● ● ● ● Path of Dick's flight

Coal fired powerplant
in operation

Coal fired powerplant
(proposed)

Coal gasification plant
(proposed)

Oil shale mining in operation

Oil shale mining (proposed)

Map showing powerplants (proposed and in operation) in the heart of national parklands, refuges, and forests complex, with coal gasification plants and oil shale stripping areas.

cent annually to the mercury concentration in Lake Powell.

Water for use by energy developers in a water-short area seems readily available, however—particularly if you have the money to buy the water rights from ranchers and farmers upstream; or have the Department of the Interior in your hip pocket politically, to help weasel away the Navajos' 45,000 acre-foot share of the Colorado for the "Navajo" power plant, owned partially by Interior's Bureau of Reclamation. The Navajos and Hopis were also clipped at the Black Mesa Strip mine where they are paid only $6.67 per acre-foot of quality well water used to slurry coal to Nevada. (Power utilities pay upstream Colorado River Basin ranchers about $30.00 per acre-foot, including pumping costs; Phoenix will similarly pay the Central Arizona project $50.00 per acre-foot.)

The 140,000 Navajos, burdened by a 65 percent unemployment rate and a population growth rate nearly twice the national average, long ago outstripped the agricultural carrying capacity of their reservation and became easy prey for industrial colonization of their lands. Yet tribal leaders today are reading the contracts negotiated with industry through Department of the Interior complicity and are realizing they were taken for another Manhattan Island ferryboat ride. Coal leases on Native American lands in the West alone will go down in history as another milestone on the Trail of Tears marking white-Indian relationships. I'd like to see the Navajos and Hopis set an example and

Loren Williams on the drag line at Black Mesa Strip Mine

seize the Black Mesa strip mine and "nationalize" it, reap all the profits from their resource, and gain control of the technical/industrial monster that is destroying their culture just when we all need to learn from it the most. The Bureau of Indian Affairs and Department of the Interior wardens have proven historically that they can't be trusted to safeguard the welfare of their "wards," particularly when big money and power politics enter the fray.

Meanwhile, political leaders in Utah, traditionally eager to pander state resources, were not to be outdone by the Navajos, and eagerly—yet quietly—offered 102,000 acre-feet of Utah's share of the Colorado to the down-river, out-of-state Kaiparowits promoters, whom the Bureau of Reclamation would charge a paltry $7.00 per acre-foot for the water, while many Utah ranchers and farmers stood ankle-deep in dust.

Utah—the new reservation.

Each power plant, coal gasification plant and oil-shale development that will use Colorado River water will also contribute to dangerously increasing the salinity level of the river, endangering valuable farmlands downstream and adding further strain to Mexican-American relations.

In 1961, the Bureau of Reclamation began dumping salty drainage water from the Wellton-Mohawk irrigation project near Yuma, Arizona into the Colorado River. The drainage water contained 6,000 parts per million (ppm) and raised the concentration of water delivered to Mexico from 800ppm in 1960 to 1,500ppm in 1962. This damaged Mexican farmlands and the resultant outcry from Mexico forced the U.S. to temporarily reduce the river's salinity until 1975 levels at Morelos Dam, where Mexico diverts irrigation water, measured 960ppm compared to 850ppm at the Imperial Dam in the U.S. U.S. treaty commitments with Mexico state that water delivered to Mexico must remain within 115 ± 30ppm of the water at Imperial Dam. Upstream oil-shale developments will raise salinity about 4ppm for every million barrels of oil produced. Kaiparowits would have initially consumed 50,000 acre feet annually and raised salinity 2ppm at Imperial Dam. Something obviously had to be done to keep the peace with Mexico in the future.

Since it would be unthinkable to ask Americans to control growth and population, let alone change their lifestyle, the beavers in the Bureau of Reclamation revealed plans for one of their more ambitious boondoggles, that Bureau of Reclamation director, Gilbert Stamm, boasted in 1975 would be the "largest desalinization plant in the world and produce 104 million gallons per day." The plant will be located near Yuma, Arizona, to treat discharge water from the Wellton-Mohawk project.

The reverse-osmosis/electrodialysis processor and related projects were estimated in 1973 to cost over $155 million. Total project annual equivalent costs for operation were estimated in 1973 to be nearly $18 million. By the time construction begins in 1979 and the plant begins operation in 1981, however, inflation will have driven these estimates out of sight. The costs will not be borne by the Wellton-Mohawk water users, however. According to a Bureau of Reclamation spokesman they are considered "non-reimbursable" by the Bureau of Reclamation and will be picked up by all taxpayers because the project helps offset the effects of upstream salinity producers—like power plants.

Ironically, the plant will consume 35 megawatts of power to operate—power drawn initially from the Navajo power plant until 1985 when a new source of power must be found. The whole project resembles a snake swallowing its own tail: the plant treats water to reduce salinity in the Colorado River caused in part by power plants that raise salinity to produce power to operate the plant that reduces salinity . . . circle game. Since the Bureau of Reclamation, Department of the Interior and their industrial cronies are blindly committed to turning the living Colorado into a urine specimen, it's easy for them to expect the unaware taxpayer to underwrite their upstream energy antics by paying for the world's biggest "dialysis" machine.

The lake loomed below us again, and Dick pointed the plane toward dome-shaped Navajo Mountain, brooding on the horizon. When we reached the mountain's foot, he turned and skimmed above an undulating landscape of petrified sand dunes. The rockscape opened up and we could see one of the most spectacular natural shrines in the world, gracefully arching over 300 feet above a canyon stream bed. It was Nonnezoshe: The Rainbow turned to a stone. A Navajo holy place.

"The Navajos consider Rainbow Bridge sacred," Dick explained to his passengers. "Yet, take a look down that way; it'll give you an idea about what we think of it. Pretty soon that water from the reservoir will back up under the bridge, and the place'll be just another tourist attraction."

Dick was right. Boat docks close to the bridge had already provided easy access to speedboaters, and transformed a trip to Rainbow from an intense pilgrimage to a casual stroll. In years past, a six-mile hike from the river or a longer overland route from Navajo Mountain had been exacted from those who journeyed to the shrine. But now the tourist industry and politicians are beside themselves with glee. The visitor head-count will soar; any kind of flotsam can now wash against the sacred bridge. Engine racket, vandalism and slothfulness are

Orienteering training at
Hunter's Point.

Ryukyuan farmer, Okinawa.

Old man, Okinawa.
Below: Demonstration at
Kadena Air Base, Okinawa.

Confluence of the Green and Colorado Rivers in
Canyonlands National Park, by Dick Smith.
Opposite: canyonlands formation known as
"Genie and the Robot".

Lili at Camp Courtney, Okinawa.
Cataract Canyon, Colorado River,
Canyonlands National Park.

now poised within easy striking distance of the Rainbow.

I found myself remembering Martin Beck of Texarkana, one of the old-time Rainbow Bridge pilgrims, who first made the long trek from Navajo Mountain to the Rainbow many years before I met him plodding along Tsegi Canyon in Navajo National Monument. We were both bound for Keet Seel ruin, and I couldn't help asking him why he carried such a huge water bag strapped to his backpack.

"Guess I got into the habit hiking into Rainbow Bridge, years ago," Martin said, explaining that the water was a real lifesaver along the hot slickrock route. Apparently Martin had found something important on his pilgrimage years before.

"I'm thinking about hiking back into Rainbow after I leave here," he mused. I playfully reminded him he could now rent a boat and ride to the bridge. My quip drew a stare like an angry god after a sinner.

"I plan to *walk* into Rainbow Bridge," Martin emphatically promised. I grinned in penance and we continued the march. Some time later, Martin lurched to a halt in the broiling sun and removed his backpack. A small spring invited us to stop a few miles from our destination. Soon its clear, bracing water poured life back in my overheated

Taos Pueblo

system. I passed a cup of its water to Martin and watched a satisfied smile spread out of his grey beard and up to his leather sweatband.

While Martin pulled his Bolex out for a few shots, I contemplated why the Hopi made sacred shrines out of their springs, and found myself wondering if an Anasazi farmer had paused at this spring 700 years before when returning from his fields to the now dead city that awaited us. Refreshed, we finished the hike and stood in awe before Keet Seel, a city of 160 rooms tucked high on a ledge in the rear of a huge sandstone cave. We were home.

We dropped our packs and rounded up the ranger who watches over the ruin. After scrambling up the ladder and wandering through the place with our guide, we mentioned to him that the moon would be full that night. Could we return and look at the ruin in the moonlight? He agreed and we returned to our camp for dinner.

It was evening and not quite dark. We were sitting in a grove of trees about 150 yards in front of the ruin with the ranger and his young daughter. Naturally, we were all expectant, waiting for the moon and the darkness. Suddenly I noticed that I was unconsciously tapping my foot to some unnoticed rhythm under the drone of our conversation. I quit tapping and asked everyone to stop talking for a moment and listen. Then I felt it again; the strange background rhythm came out of the silence from the trees overhead. The others cooperated by remaining silent but didn't immediately understand what I was listening for.

"What do you hear?" I asked.

"Crickets," someone answered.

"Listen carefully." I stood up in a stooped position and began a chant I had heard at LaSalle's place, followed with some dance steps I vaguely remembered. The crickets kept perfect cadence chirping in unison. It grew silent then and I watched realization spread across their faces. It was a strong coincidence at least. At most, the night animals' sounds echoed a primeval model from which ancient drummers may have taken their first cadences.

The ranger was quick with a rational observation which disturbed the spell slightly as we agreed with him that, "yes, the cadence was often directed by the temperature," growing slower when it grew colder. As we prepared to move up in the moonlight up into the city, the crickets serenaded us like a thousand drummer spirits, preparing us for our entrance to the kiva. It was time to go up.

The kiva is traditionally an underground ceremonial chamber that symbolizes the womb of the Earth Mother for Pueblo Indians. As I settled along the back wall of the kiva, the full moon crested the opposite cliff rim and flooded the cave with persistent, hueless light. I jumped to my feet and gazed at the deserted city—a dusty monochrome from another time. Martin sat cross-legged on a roof top, alone with his thoughts, gazing into the canyon below. The moon climbed higher, perfectly centered under the soaring arch of the cave's roof, directly above the kiva. I sat down to think.

In times past, the young boys of the village entered this kiva when it was time for them to become men. There they stayed and learned the secrets of their religion, which was life itself for them. The history of their people and the sufferings they had endured when they closed their hearts to the truth were revealed to the initiates. But more importantly, they learned about interrelationships and their role in the universe as part of one living whole. They learned to join with other living creatures, not live apart from the rest of nature.

They were tied to the seasons and cycles, and the richness of their rituals provided symbolic bridges to their unconscious roots, helping to keep them whole. And they learned to move carefully and deliberately in their actions. Survival in a harsh, marginal environment left little room for error.

Taking the life of another living being—a tree, a deer—was only done after careful examination of one's motives and needs. Even then, the creature's permission had to be asked in prayer, accompanied by a promise that the sacrifice of the victim would not be in vain, but would improve the capability of the petitioner to someday return more than he had taken.

Small wonder that many of the great Pueblo cities of the past had been occupied longer than the United States has existed. Keet Seel, Pueblo Bonito, Cliff Palace—a few of many silent reminders of a great people who flourished in the desert vastness of the Four Corners region, only to move on and disappear.

Yet their descendents linger on in Hopi villages like 700-year-old Oraibi, the oldest continually inhabited city in the United States.

When the young men had emerged from this kiva, they were considered literally reborn from this earth-womb, no longer owing primary allegiance to their human mothers. Instead, they were sons of Mother Earth; brothers of all her creatures. It was an awesome responsibility to bear, but they were now men and prepared to undertake the journey.

I looked around at the deserted city that night and wondered what had brought its inhabitants to their downfall. Had they lost their topsoil or been driven out by drought? Perhaps they had denuded their forests for a distance so far that they could not economically gather fuel for cooking and heat. Or maybe they had neglected their defenses and become so specialized and dependent on agriculture that they

The kiva and walls at Chaco Canyon

had to watch in helpless rage as lean, hungry raiders plundered their fields. Thoughts of men carrying AK-47 rifles and bandoliers of rice came to mind—modern raiders hitting the jugular of a highly specialized nation that consumes 30 percent of the world's resources and has lost control of its appetites.

In the middle of this dead city, the conclusions seemed inevitable. Particularly if you try to think like a mountain, or even better a river. *We need to grow with less, and move toward a steady-state or maintenance economy.* The countryside is fast reaching the limits of its capacity to support the giant megalopolises; ultimately they must break down and people must return to support the earth.

Martin and I climbed down the ladder and returned to camp. I put on my jacket and lay down in front of the fire to sleep while he rummaged around inside the little tent he had lugged along.

"Smith, aren't you a little cold without a sleeping bag?"

I made some noncommittal reply, feeling it wasn't necessary to point out that the crickets' tune had slowed to a largo.

"Here, take this. It'll help keep the chill off." Martin tossed something over in the dark. I chuckled a little hypocritically, and then curled up in a thoroughly modern, technological miracle called a Space Blanket.

"Crickets, eat your hearts out."

"How much farther is Grand Canyon?" Nadia Stolt-Nielson called up to Dick, as we cruised a couple of hundred feet off the deck.

"Ohh, not too far," Dick grinned, "I thought we'd let you first see it a little differently."

Suddenly we rocketed off the North Rim of the Canyon and the earth dropped away below. Nadia gave a squeal of delight. It felt like the whole world had been snatched from under us. Dumbfounded, we stared as the evening sunset draped the gorge below in varying hues of rose, purple and burgundy. It was a fitting climax to a perfect flight.

It wasn't the end, however. Dick was scheming again. He flew the canyon sunset like he was directing an orchestra; the Stolt-Nielsens clapped for encores, as he adjusted his course to catch the best lighting effects for his guests.

"You see the river down there," he shouted. "The Grand Canyon has the best rapids left on the Colorado. Hope you can run them sometime. Now where do you folks want to go?"

"Go now? I thought we were going here?" I knew he had something up his sleeve, and I felt a second-hand seizure, or clock attack, coming on. Sure enough, the Stolt-Nielsens decided they couldn't "see" the

Grand Canyon any better than this sunset service and were open for suggestions.

"Well now, how 'bout Las Vegas." I caught the sarcasm in Dick's expression, but brushed it off as he changed course for Sin City.

"Is Dick always this impulsive?" asked Nadia.

"No he's usually a lot worse," I teased. "He's kind of quiet today. Usually he'd head for Yucatan."

It was nearly dark when we first saw the lights of Vegas glinting out of the somber browns and greys of the desert below. As we grew closer, the blare of lights became an electronic fever blister.

"The damn place looks like a blight," I shouted. "I feel sick . . ."

"Wait'll you see the topless dancers," Dick tossed out.

We landed, bid farewell to the Stolt-Nielsens, and spent the rest of the night ambling in and out of casinos. Dick treated me to a show at the Stardust; I bought him a Japanese dinner to flaunt my new-found knowledge of the Orient. After about three days, the sun came up next morning and we crawled sleepily into the plane and somehow got her pointed home. We had met the enemy and *he really was us*.

"Hey Dick, were you tryin' to tell me something on that flight?"

"Well, maybe. There's plenty that needs bein' done. What're you doin' after the Marines turn you loose?"

"I don't know; what do you have in mind?"

"Well there's always the Park Service; they need all the help they can get. And there's the new Maze District across the river. They haven't managed to ruin it yet. Maybe you can get out there before it goes under."

"Thanks for the overview, Dick."

For some reason I hadn't gone flying with Dick that weekend. Instead I'd left the little Maze district ranger station when thousands of migrating butterflies were fluttering over Robbers Roost, and driven to Logan, Utah, to see Lili.

When I returned a few days later, the butterflies were gone . . . so was Dick.

It was as if the unseen force or spirit that breathes life into the immense canyon country had whisked both Dick and the butterflies away toward that unfathomable journey to the divine milieu.

Dick Smith was killed in a plane crash on May 19, 1973 in his beloved Canyonlands. He had been searching for Desert Bighorns with three Park Service employees along the Colorado River. The crash came two months after I'd taken his advice to work in the Maze District.

It hadn't mattered what remote airstrip we touched down at;

somebody always called Dick by name. "How's it goin'?" he'd grin back. His generosity was legendary. At his funeral one of his longtime Monticello neighbors eulogized Dick's passing by recounting the countless times someone had needed an emergency medical flight to Salt Lake—often at night, in marginal weather. "I'll be there in twenty minutes," was always Dick's reply.

Dick was born in Monticello, Utah, where his parents, Jo and Kip, run the Wayside Motel. Active and extremely independent, he tried college for a while, but flying was always his first love. He'd been a crop-duster, deputy sheriff, truck driver, cat skinner, pilot and resort manager. As a boy he had trapped and learned to live off the land. It always seemed like Dick tried to cram as much living as possible into his double-time days. He exulted in living close to the cutting edge of life. He was complex and sensitive, loved music and photography. But most of all, he loved the canyons. He was thirty-five when they claimed him.

His death left an incredible void in my life, and roaming the twisting canyons of the Maze country seemed a little empty at times. The months following Dick's accident were beautiful, however. Days took on added intensity and sometimes, in the silence of the backcountry, sudden discoveries or the peace of being one with the land made it seem almost as if Dick were directing my wanderings. The years I

Dick Smith

spent in the Marine Corps were a drought as far as songs were concerned. I'd managed only two melodies in more than three years, and words had come for only one of them, "Gypsy of My Mind." The other was "Hey, What's Happenin'?"

The Maze therapy that Dick prescribed was the right medicine. Songs began to surface again and the idea for this book came too. After the season ended, I prepared the first chapters for presentation to publishers. Then, just minutes before I was scheduled for a recording session to make a demo tape, I was tuning the guitar when the old melody to "Hey, What's Happenin'?" came back and words started to tumble out. Lili grabbed a pencil and paper and wrote them down as I recited. After a few minutes we stared in amazement at the paper. The song was fulfilled—a farewell to Dick.

My thoughts return to the evening I was driving a park service rig back to the ranger station after Dick's funeral. After a while I couldn't take it any longer, parked the infernal clattering machine, and wandered across the silent Roost toward the distant Henry Mountains. It was late evening and the slant-light of the sun cast a glow over the translucent spring greenery.

Suddenly I saw them. There were three or four of them. Back-lit in the golden light was a lone group of butterflies hurrying across the flats, pulled by some magnet-like force.

"Stragglers trying to catch up," I thought at first. But as I stopped to watch them disappear, the beauty of the scene reached down through the sorrow of the moment, and I began to wonder . . .

Hey, What's Happenin'?

Hey, what's happenin',
In your tri-color double-time day?
Are you still weavin' ribbons through an orange-colored sky,
As you unwind your way?

Well I dig down through my ragged brown rucksack full of
      mem'ries.
You remember, the ones you helped to store away?
And I ask ''Why the future's always early,
And tomorrow comes before we've really lived enough today?''

I wonder if you're someplace where you can see me,
As I hike along this stream bed and kick a rollin' stone.
Yes, I miss all the good times we shared together,
Still I know good times are bound to come just like ours were
      bound to go.

And I see you're movin' on the clouds now,
Silver jet-stream takin' you away.
While the canyons fill with storms and cry with moaning
      thunder,
And the raindrops form the tears that part the years and carve
      the days.

And I see you're movin' on the clouds now . . .
Silver jet-stream takin' you away . . .

14. Birds Fly
Around Her

The Maze

The young ones were beautiful. I watched them rise out of the desert like a vision of hope. Through the dusty, golden cross-light of the sunset, I saw the serpentine line of tired marchers moving over and through sandy folds of the barren landscape toward our rendez-vous point near the Dirty Devil River in Southern Utah.

If I squinted my eyes as the struggling figures grew larger through my blurred vision, I was propelled back in time—far back—and my imagination forged an image of those who first wandered onto the continent seeking new homelands. According to the Hopi, as re-corded by Frank Waters in *Book of the Hopi,* they were greeted by Masaw, the protector, guardian, and caretaker of the Fourth World. He instructed the new arrivals to purify themselves by first wandering in the Four Great Directions to know the land intimately, and make themselves spiritually worthy of claiming it for the Creator—Taiowa. Masaw warned them not to succumb to greed and materialism by settling in lush lands of plenty and forsaking their migrations. And he urged them to stay attuned to the promptings of the universe by tuning themselves to the vibrations of the land they were consecrating. Vibrations protected and sustained in the living bodies of earth and man by the twin gods, Poqanghoya and Palongawhoya, who reside respectively at the North and South Poles, jointly keeping the world properly rotating and sending vibrations along the earth's axis through its vibratory centers.

Palongawhoya—the Echo Twin—controls the wind's movement, and sings warnings through these centers. The People, then, catch these promptings (songs?) through their own vibratory centers (unconscious?) like the kopavi or soft spot in the top of their heads that serves as an "open door" or drumhead to the vibrations of the universe. When the People harden their hearts to the earth's vibrations and close spiritual doors to the Creator, they are punished; they succumb to their own lusts, greed and materialism—madness.

Finally, Masaw warned that if they became corrupted by evil and their appetites, he would take earth away from them.

So in keeping with this grand act of *attune-ment*, the People wandered in the Four Great Directions leaving evidence of their passing, like Kokopilau—the Humpbacked Flute Player—whose likeness is carved in rocks as petroglyphs from the Canadian border to the end of South America. Kokopilau—symbol of the Hopi Flute Clan—the happy backpacker carrying seeds for plants and flowers, bringing warmth through the breath of his flute's music, "the gentle, ordered movement of wind."

Their wanderings resembled a giant cross on the North and South American continents, with its center located south of the area now known as Four Corners, where the States of Colorado, Utah, Arizona, and New Mexico join. Here are the Hopi mesas where the arms of the migrations came together: Tuwanasavi—Center of the Universe. Home.

I opened my eyes and the scene snapped back into focus as the modern pilgrims came near. These young people were involved in a migration that would help them become "native Americans." They were survival students from Brigham Young University, and were just finishing the first third of a month-long, 250-mile trek through some of the most rugged country in the world.

Later that day, I watched the fifty participants disperse in cooking groups and set to work trying to coax fire from flint and steel sparked into dry juniper bark. Then they pooled their meager rations of water, flour, grain, raisins, oatmeal and boullion cubes to make a meal of lumpy-dick soup, ash-cakes, and cracked wheat.

"Would you like to share our food?" a student asked. I felt a little guilty, but did anyway. A familiar face grinned across one of the cooking fires. A few days before she had earnestly asked me to watch for a canteen she'd lost on a night march. It belonged to her brother and had "heirloom value," as she had put it. I mused about changes in priorities in the desert. A canteen becomes one's most important possession, roast grain and oatmeal a feast.

When dinner was finished, we gathered around the fire to share songs and experiences. The night passed quickly. We were singing "Dick's Song" when the Pleiades rose above the horizon; it was time for me to leave them. They still had thirteen miles to go that night toward the Henry Mountains—the last mountain range in the United States to be named—where they would pause for five days, each participant going off alone to experience complete solitude. It would be like the "starvings" or "name-seekings" of Native Americans, when young people enter the mountains to fast, pray and seek identity. I turned toward the Henrys and noticed high cirrus clouds etched across the face of the moon. "Storm coming," I thought, "the kids might get cold since they only have one blanket each." Then I smiled quietly to myself, remembering the times darkness had caught me miles down some intriguing canyon without a blanket. They'd be okay.

Probably the greatest joy I had working as a back-country ranger in the Maze District—aside from the self-indulgent pleasure of being able to explore such wonderful country—was sharing moments with groups like the survival students, and the people who were willing to come to a remote area like this and bend to the country,

''Nuts and Bolts,'' near the Maze, canyonlands

rather than try to make it bend to them. Our operation at that time was informal, relaxed and flexible. We spent our days ranging, and the folks we shared with were people, not clicks on a turnstyle.

While the rest of Canyonlands provided a spectrum of recreational opportunities for visitors—from the windshield tourists and four-wheel drivers on up—I felt the greatest asset of the Maze District was its feeling of remoteness or removal. It was something to be earned, something to prepare for, demanding greater levels of skill and understanding. This was no place for intensive development, bales of brochures or asphalt trail systems. It was a place where the possibilities for feeling self-discovery among those who visited should be protected as long as possible. Let them feel they might be the first to discover a pictograph, new route or spring; and make sure there was enough country left where they *would* be the first to stumble on some exciting discovery. The people came. All ages, backgrounds, and income levels, drawn by something intangible, still unspoiled.

In the high school at Green River, Utah, I found many of the kids had never visited our part of the country, even though it was in their own backyard, just seventy miles away. Yet, after hearing the songs the country was inspiring, and after seeing slides of what awaited them, many were ready to come. I'd been invited to speak to the students by their literature teacher, Linda Findlay. Linda was an intellectual oasis in this tiny desert community. Her students were reading stuff I hadn't been exposed to until college, and they showed promising signs of developing a healthy, balanced environmental ethic. Linda hoped to introduce environmental education into the Green River school curriculum, utilizing the park as a wilderness laboratory, and many of her students seemed anxious to be involved.

The youth are the ones who will find creative ways someday to reestablish contact with our roots and become good husbanders of the earth. This was demonstrated for me one day when I stopped at an oil well site outside the park after talking at the high school. A big, burly student of Linda's named Rodney grinned at me through a grimy face. He'd been skinnin' his father's Caterpillar and wanted to show me his handiwork. I shuddered a little bit; some Cat drivers really tear the hell out of the country. Not Rodney, though. He cocked his hard hat and started to explain.

"You see how I did this, Ranger Smith? Instead of just comin' in here and tearin' out all those pinyons and junipers, I convinced them to split the site in two." Rodney proudly continued, "By doin' that, I was able to put two small sites in where fewer trees would be hurt." He was right. He had managed to locate the entire drilling unit with minimal damage.

"Hope you notice something else," Rodney added. "I didn't make big turns and sweeps with the Cat around the site. You run over fewer trees that way." I wanted to show Rodney I appreciated his efforts. I was headed for the Maze, planning to hike to some pictographs called the Harvest Scene. I knew he wanted to see them, and asked if he'd like to come along.

"Well, how long will it take?" Rodney fretted about wasting time, he was a workin' man, you know. The zealot in me came out. I teased Rodney about worrying about clocks in a place like the Maze.

"Now, Rodney, you know what I told you about comin' to the country on her terms, not yours. Remember, you have to bend to the country; you can't make the country bend to you!" The Cat driver in Rodney came out and he quickly slammed me back down to earth with a sly grin, saying, "Yeah, unless you have a Cat-er-pil-lar!"

Rodney threw a cooler full of Mountain Dew in the back of my truck and we were off. Soon we scrambled down a cliff into the Maze and hiked through its hot canyons to the mysterious paintings. Rodney stood in awe before them. Huge painted figures, beautifully rendered with exquisite detail, proclaim a thousand-year-old statement on ecology.

One of the figures with arm outstretched offers a "tree of life" to the heavens. Actually, it is a bunch of rice grass—an important staple of the people who roamed these canyons for centuries. Flying toward the offered rice grass is a large bird. Rabbits run down the man's arm; a coyote waits behind. Below, harvesters stoop with a wooden seed-beater and a sickle, fashioned from the horn of a desert bighorn. The figures are frozen inextricably together for all time—interdependent, whole. Rodney watched them under the shade of his hard hat, slurping spring water from a Mountain Dew can.

The paintings and petroglyphs scattered throughout the canyonlands, particularly in western canyonlands, hint at many stories. Undoubtedly, the Maze District contains some of the finest primitive art in the world. One panel of paintings in mysterious Horseshoe Canyon alone stretches along an overhanging wall for more than 200 feet. Many of its figures are life-sized humanoids that stare back at their beholders with open, haunting eyes. Bighorn-sheep hunters prepare to spear a band of their prey, while a dog and running youngster drive them within range. A ghostly procession of marchers, carrying packs on their backs, moves along another wall, looking strangely like the lines of modern-day wanderers I've seen threading their way across the desert.

As I explored more and more of the Maze country, I began to find ancient travel routes skillfully interconnecting the canyons, passing springs, flint chipping sites and granaries, and primitive shelters in caves and beneath overhangs. Many canyon walls seemed at first unscalable, only to yield to a climber because of tiny, inconspicuous toeholds that had been strategically scraped in the rock face, then slowly encrusted with lichens after centuries of disuse.

Jasper Canyon—sometimes called Defiance Canyon—is well endowed with examples of such holds. One precipitous friction pitch in particular looked quite formidable at first, until a closer look revealed a series of toeholds discreetly carved up its steep face. Once up the pitch I was stymied by an impassable overhang.

"What gives?" I thought. "This is the way to water from above, all right, but why the dead end?" I sat down to rest on the lip of the ledge and dangled my feet over the edge. Morning light crept down between the deep sandstone walls of the narrow canyon. The canyon wrens sang a blissful accompaniment to the light's arrival—sounds like silver icicles cascading down sandstone walls. The moment called for a toast.

I mixed a packet of lemonade with the water in my quart bottle—water drawn earlier from a spring now illuminated below. After a mouth-puckering swig, I watched the sunlight glint off the flakes of jasper that blanketed the ground near the spring, broken arrowheads and imperfect stone tools cast casually aside by warriors and hunters who had paused near this oasis hundred of years before to replenish their armory.

My morning reverie was interrupted by an aerial intruder that floated eerily out from above the overhang to give me the once-over. A buzzard. While patrolling the cliff rims for a possible breakfast, my friendly visitor had noticed the strange morsel perched on the ledge below.

"Not a bad way to go," I thought. Being transformed into buzzard flesh to soar the canyons has a lot more appeal than being shot full of chemicals and walled up in a concrete vault to wait for the worms of year 4024 to finally batter their way to my remains. Glancing beneath my feet at the steep face and footholds the ancients once scaled with impunity made me curse the sedentary blight that affects my race and wonder if the old buzzard might have his meal sooner than expected. Unless I could find the answer to how the old ones had scaled that overhang, he just might. Trying to descend the face didn't seem appealing at all.

Then the answer came. Lying behind a large boulder, under the protective overhang, were some extremely old poles that had been cut with stone axes. The wood was well preserved by the desert dryness and looked sound. I raised one up and laid it against the lip of the

Millard Canyon and Wolverton's Cabin, Orange Cliffs, Utah

overhang. It fit perfectly in a small notch I had failed to notice earlier. I shinnied up it and in a few seconds was on top of the cliff, blinking in the warm morning sunlight.

As more and more of these travel routes unfolded in the Maze District it became apparent that this section of the park lent itself ideally to backpackers—pedestrians following the paths of others long since gone. With a knowledge of some of these routes and water points, a hiker carrying a quart or two of water could traverse the area quickly and quietly, moving through some of the most extraordinary and revealing terrain in the world. There in the undisturbed silence of this small corner of Canyonlands National Park, modern sojourners could place their feet and minds on the Beauty Way of the ancients.

A short time later, I began to picture why and how the old hikers may have traveled these routes. Were some of the travelers who frequented the Maze District Anasazi from the plentiful cliff dwellings on the eastern, Needles side of the river? Were they on harvesting and gathering expeditions, supplementing their farming efforts in Needles with rice-grass seeds, pinyon nuts, sunflower seeds, yucca plants, prickly-pear fruit or sego lily bulbs? I imagined lines of harvesters collecting food at the lower elevations first, then climbing to higher levels as the plants progressively ripened uphill, finally cresting on the uppermost level, the Orange Cliffs, just as the pinyon nuts were ready. Or perhaps they came to restock their armory and stone tool supplies from jasper deposits and flint beds.

A huge boulder perched high on a ledge above the east river bank and covered with petroglyphs seemed to support these conjectures. One intriguing scene shows two adults holding a child between them by his outstretched arms. Were they proud parents, showing off their child, or were they fording the Colorado near the Spanish Trail, keeping the child's head above water? Lines of marchers carrying backpacks moved silently past the parents, across the rock and toward the river. The rock also held an echo from a more recent traveler. A faint signature, considered by some to have been left by French fur trapper Denis Julien in the 1830s when he explored Cataract Canyon three decades before John Wesley Powell, spoke from this ancient register.

There were few extensive cliff dwellings or pueblo ruins on the Maze side of the river, yet there are many strategic granaries hidden along the lines of march that still have well-defined fingerprints of their makers imbedded in the mortar. Did they cache excess harvests for those who would follow the next year? Those who made the inconspicuous toeholds in the canyon walls either moved in stealth or were lazy. Or maybe they were just darn good rock climbers.

Perhaps the Fremont Indians shared or made the routes. Some experts believe most of the paintings in western canyonlands belong to the Fremont culture. However, the Anasazi definitely used the routes I followed, because broken bits of Anasazi pottery littered both the routes and the springs. One dark recess even contained a perfect, small black-on-white bowl inside, once used to quench the thirst of an Anasazi traveler, as it did mine.

The bowl is still there; maybe you can find it. If you do, leave it alone. Carry it away in your memory only. There are many other seekers yet to come, in need of refreshment. And if you don't mind, when you find the old pole I left at the overhang, drop it down; I forgot to put it back under the protective ledge.

BIRDS FLY AROUND HER

In three widely separated locations—Horseshoe Canyon's Great Gallery, Spur Canyon's Blue-Eyed Princess, and the Maze's Harvest Scene—I found the same theme repeated in the wall paintings. In two or three places, human figures wore a "white crown," perhaps a garland of prickly-pear or yucca flowers, and birds were hovering around the figures—mostly hummingbirds.

Hummingbirds, sacred to the ancient Mayas, Toltecs and Aztecs, were symbols of everlasting life. Aztec leader Moctezuma reportedly had a crown adorned with their figures, and one of the chief deities of the Aztecs, Huitzilopochtli, was known as the Hummingbird Wizard. When the Aztecs wandered into Anahuac, the valley of Mexico, around 1168 A.D., they brought the idol of the Hummingbird Wizard with them. During their travels they had consulted it for advice, and had been instructed to prepare for migrations that would lead them to new homelands by sending ahead smaller bands who would plant crops for the main group's arrival during the harvest. The Wizard also demanded a grisly diet of offerings—human hearts freshly torn from their victims.

I've often wondered about the Aztecs, who were linked linguistically to some groups of Indians like the Hopi, Commanche, Pima and Shoshoni of the Uto-Aztecan stock, and pondered whether these landless wanderers, who entered Mexico around the time many southwest pueblos were being abandoned, might be directly related to the peoples of the Great Basin or Four Corners region. Vigorous trade was definitely carried on, with copper bells, jewelry, macaws, parrots, etc., being carried well into Utah. A parrot-feather mask was even found in Canyonlands National Park in a canyon once inhabited by the Pueblo Indians. Mesa Verde culture is commonly called

Anasazi, which is a Navajo word meaning "Ancient ones."

The feelings I gathered from the hummingbird paintings and the figure they hover around were far more gentle than those suggested by the insatiable Aztec god, however. The image of ancient travelers stopping to record the recurrent theme of the hummingbirds was captivating indeed. Was it a thousand-year-old statement of everlasting love, as well as everlasting life? Was the garland-clad figure a girl? If so, she must have been a remarkable person, and her admirer one helluva hiker, because at least two similar paintings are located as far apart as White Canyon near Hite, Utah, and west of where Interstate 70 cuts through the San Rafael Swell—over 120 miles as the crow flies!

"Birds Fly Around Her" came to me in a roundabout way. I had stopped to camp one night far out in the Navajo Reservation, east of the volcanic plug, Agathln. It was sunset and a warm, steady wind streamed toward my campsite from the west. I picked up my guitar to salute Father Sun as he nodded out of sight, and the old melody of the Sawtooth "Mystery Song" came, incessant and repeating. I played it over and over, while new words and ideas sprang forth; marvelous variations held me in a trance.

The flat expanse I wandered on vibrated like a huge, living drumhead! The wind felt supercharged with electricity, and I was oblivious to everything except the music that bound, yet swept around and through me. The long-awaited song seemed trapped inside, straining to burst through the top of my skull. But it didn't. Instead, I came to, standing a half-mile from my truck. Shaken, I walked back to camp.

It had been a long time since I'd caught a new song, but a few days later, the song drought ended, broken by this experience and Dick's "Maze remedy." Although the "Mystery Song" resubmerged, "Birds Fly Around Her" broke free while Ranger Victor Carmichael, Student Conservation Aide Karl Von Berg and I were sitting under the pinyons near my small trailer house at the Hans Flat Station. Karl played harmonica, I picked guitar. When the words and melody started to tumble out, Vic grabbed a pad and pencil to catch it. We all looked at the new arrival on paper. It had been a long dry spell.

Birds Fly Around Her

Deep among the canyon walls, I hiked alone while silence flowed;
In some cool shade I stopped to rest and drink some water.
Looking up, I noticed her, an act of love was painted where
The sandstone face slipped down to meet
        –the crumbled ages.

Unknown hands from years gone by, had stopped to rest here,
        the
        –sume as I;
But with his brush and paints of clay, he left his love here.
A painted girl to greet the dawn.
A crown of white, a feathered gown;
Her smile demure, she waited while
        –birds flew around her.

A thousand years this love has grown
        and stood the test of wind and storms,
While my love waits beyond these walls,
        –birds fly around her.

The canyons grow and canyons die, as sand dissolves behind her
        eyes,
Her painted birds will tumble down and join the ages.
The grass turns green and then to brown, a thousand times
        before her crown;
Although she's changing, slowly fading,
        –her love will live here.

So who am I to linger here?
I'm just a flash before her mirror,
A fading whisper who stops to watch
        –birds fly around her.

15. The Red Tail

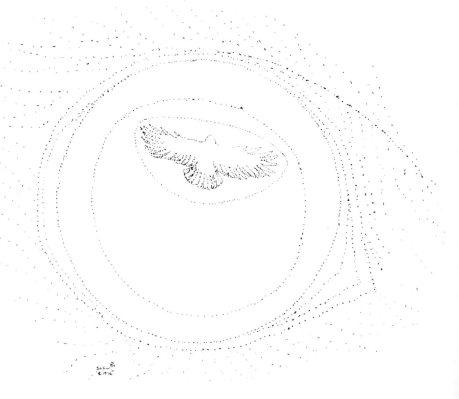

During my last season in the Maze District, Ranger Doug Treadway, Student Conservation Aide Cliff Casperson, Cliff's father, maintenance man Steve Riley, and I were hiking down to a huge alcove we call the Concert Hall. A spring-fed, icy-clear pool sits in the bottom of the chamber under a 200-foot overhang, surrounded by white sand and horsetails. Cool, moist shade helps support the river birch that grace this green mansion in the desert.

A sandstone projection extends from the overhang above the alcove. Here it has always been a custom to make music, echoing songs into the giant chamber, where offerings of voice, flute, and guitar would be gathered and shaped in the desert silence.

As we passed along some sculptured sandstone bathtubs above the Concert Hall that night, a red tail hawk suddenly flew up in front of Cliff and me. Wings outstretched, he screeched his call and lifted into the evening sky. As he circled above, the slanting rays of the sun puffed his tail transparent red. Treadway and the rest stood openmouthed. It was a good sign. The hawk flew out of sight over the ridge above. We continued around to the projection and surprised another red tail—the mate—making her own music. She pitched off the cliff, soared into the void, and rose above us, where she continued to circle. We sat down to watch, and I pulled the guitar off my back.

While watching the rhythm of the hawk, I began to pluck the strings and whistle a strange melody—a melody that grew with her flight. Sometime later, she flew off. Treadway turned with an inquisitive look.

"That's a nice song," he said. "Where did you learn that?"

I told him I didn't know, and half-jokingly said, "The hawk must have taught it to me."

He grinned, "Play it again."

"OK." I couldn't. The song had gone as fast as it had appeared.

Many weeks later, I injured my back on a raft in Cataract Canyon and was slapped in the Moab hospital. While convalescing, and to keep from going nuts in the confines of the city, I'd sometimes get an airplane and fly out to watch the sunrises and sunsets over the canyons. One morning was particularly beautiful. The sunrise burst through hanging veils of scattered thunderstorms, turning them into a brilliant metallic gold: "Sunrise, a burst of glory in the sky . . ." the words had just come.

Later that night I retired alone to the bank of a small lake. While playing the guitar in the darkness, I suddenly found myself engulfed in the hawk's melody again. I jumped in the car and drove like hell to Arches National Park, where I rousted Ranger Jim Martin out of the rack.

Brian Alberg and Patricia Udall at the windbreak

"Get a pencil, Jim Bo. Something's coming." Jim scribbled furiously and soon we had the song the hawk had generously decided to share again. The next day I strapped my guitar to my backpack, and clambered down the cliffs to where a bunch of Student Conservation Association kids were camped in the Maze. I sang "The Red Tail" with them that night. We sang "The Red Tail" the next morning. We sang "The Red Tail" again and again. The song fit.

The Red Tail

I am the Red Tail
Climbin' and soarin' through your sky,
With the sunset on my feathers,
With your friends, all gettin' high!

Higher than the Red Tail,
Up above me, there is beauty all around.
Out beyond me, there breathes beauty.
Down below me, there grows beauty.
There is beauty all around me!

Learn to see me,
Learn to feel me, like the wind across my wings.
Let my spirit grow within you.
Learn to know me,
Learn to be!

Like the Red Tail,
Catching currents, and rising in the sky.
Out away from all that's ugly,
Breathing freedom from the windstorms,
Growing wise and filled with light!

You can be the Red Tail
A sunrise, a burst of glory in the sky.
You'll know freedom,
You'll know beauty,
You'll find love, and you'll be wise!

Rise with the Red Tail!
Strive to find, all the beauty in your Life.
Like the Red Tail you caught at sunset,
Like the Red Tail of your mind!

I am the Red Tail,
Climbin' and soarin' through your sky,
With the sunset on my feathers,
With your friends all gettin' high . . . all gettin' high . . .
all gettin' high.

16. The Mystery Song

A Native Son

Ron Warnick, a dairy farmer from Montana, is one of many young Americans who have been and are being shaped by the rhythms of the American earth. Slowly, as it must have happened with the first Native Americans, the spirit of the continent is permeating its inhabitants with a sense of place and community. The most encouraging thing of all is that many Americans have listened to the warnings of the unconscious, and are willing to risk the honesty and sacrifice necessary to come to the earth on its terms, to learn to live with it and its limitations, and to recognize that true progress is creatively providing for our needs with less, not more.

The wilderness has been an important factor in shaping Ron Warnick and others like him. Ron spent many years roaming the North American wildlands from Alaska to Mexico, once living entirely off the land for a summer in the Bob Marshall Wilderness Area. Along with his mother, Donna, and wife, Lou, he fashions buckskin clothing hand-laced with leather and fastened with antler buttons. My finest, most practical and comfortable piece of clothing is a buckskin jacket Donna made in appreciation for a trip to the Maze and some songs I shared with them.

During that Maze trip, I took Ron and his mother to a special place I'd discovered earlier in the year. I'd accidentally stumbled upon an overlook of the canyonlands while trying to pinpoint the location of a fire that was burning in a bottom along the Green River. I had left my power wagon and run cross-country for about a mile until I reached a spot along the cliff edge where I could see the river far below. The fire was burning safely in some tamarisk, walled in by the river and sandstone cliffs. It became of secondary importance, because I'd been struck by the staggering grandeur of the scene that unfolded around me.

Before me loomed the Buttes of the Cross, named by John Wesley Powell during his exploration of the Green and Colorado Rivers. My perch was about level with their flat tops, and we were separated by the golden folds of the lower reaches of Millard Canyon that undulated a dizzy 1,600 feet below. The Green River slipped lazily through the meanders it had carved in the white rim sandstone. A prairie falcon rocketed along the cliff in a humbling display of aerobatics as I watched enviously from my perch. The air ripping around the falcon's wings was the only sound to disturb the perfect desert stillness that day; even the wind was quiet. The pillar of smoke rose straight to the cerulean sky. Everything felt right. Complete.

When Ron and Donna hiked out onto the knob with me, they

were silent. I was pleased to watch them withdraw into themselves
and the sanctity of the place. While the buttes and desert floor beyond flowed with the changing reds, oranges and purples of sunset,
a dark shadow crept across a sandstone wall facing us. Behind us a
large cove receded, exposing an exquisite tapestry wall filigreed with
glistening stains of desert varnish. My attention was focused mostly
on the changing colors in front of me, but occasionally I'd catch
Donna and Ron out of the corner of my eye curiously looking at me
in the silence. I didn't give it much thought at the time; figured they
were just quietly expressing thanks for the moment.

"Well, did you see it?" Donna couldn't contain herself any longer
and was the first to break the silence after we returned to the truck and
went rattling toward the ranger station.

"What do you mean?"

Ron laughed, "Yeah, I saw it too. It's just as ugly as you are." I
took the teasing but wanted to know what they'd seen that I hadn't.
They were incredulous.

"You mean you've gone out there before and never seen yourself
grow old and die?" Ron asked. "As the sun was setting, your profile was
in that big shadow that moved across the wall!"

"Oh bullshit. You guys are putting me on."

"No. It's true," Donna seconded. "And as it moved, you got

The Warnick family

older. Your hairline receded, your nose became hooked, and your gums or mouth looked like an old man's. Then you disappeared."

I began to wonder. They couldn't have made it up. There hadn't been a chance for them to plan it. I wavered, and said I hadn't seen it.

"Well, it's you, all right," Ron grinned. "The exact same profile, big nose and all. Go out there again and see for yourself."

A few nights later, I wandered out to the spot alone, carrying my guitar, hoping to deal with "The Mystery Song" while waiting for the sun to go down. The song's melody fit the place perfectly, like it fit other natural settings. And as usual, I wasn't ready to catch all the words. Just a few.

As the sun lowered behind me I watched the wall. Then it came, just as the Warnicks had described it. A man's profile was silhouetted in the shadow that moved across the wall. And, it did change, as if aging, until the shadow engulfed the cliff and melted into the night. I sat for a long time, not moving, hardly breathing.

There *was* a disturbing likeness. As the darkness gathered around that little promontory, I began to understand what "The Mystery Song" meant. The place was like a time machine. Not only did I see a projection of my own death, but I had my song to go with that time, whenever it came, like LaSalle Pocatello's secret song.

I was strangely comforted and ready to let the melody recirculate

The Tapestry Wall, in the canyonlands

through my life according to its own law. Feeling a little chilled, yet quietly exhilarated, I picked up the guitar and walked through the darkness back to the truck.

I've thought of that evening a great deal recently. After I had finished the manuscript for this book, I suddenly became ill and lost my ability to read and walk normally. The doctors say I have multiple sclerosis, and although my reading and walking ability might improve, I know now my life will be changed. It's difficult to say how many more trails will feel my boots upon them.

As I prepare to face this new adventure, I am deeply indebted to those in our country who had the courage and wisdom to preserve the areas of wildness in our midst that helped me grow. Just the knowledge that those areas, and hopefully many more, will always exist gives me strength and hope, especially when I think of the young ones yet to come who might have the opportunity to experience what I did, and to be shaped by the American earth. Our wildland legacy will undoubtedly go down in history as the finest hallmark of the North American civilization.

I asked Ron Warnick to pen some of his thoughts for this book, and his words capture the spirit of the many native Americans I've met and grown to respect over the years:

"It begins in darkness.

"The wind blows about us, blowing sounds of unseen leaves and grass and swirling clouds whose presence we feel but only dimly see. The first light is a deep blue, born out of blackness.

"The clouds and the land grow and solidify in a kaleidoscope of color. The second light is orange-red, which quickly expands into the yellow brightness of dawn and then the full white light of day, as the sun emerges over the faraway highwood mountains.

"We stand, my wife and I, facing the sunrise. For there is power in the rising sun.

"Behind us is the house—our home—which we have fashioned with our hands out of wood and stone. The supporting poles sink, like tree roots, probing our mother—the earth. The house rises low and solid against the prairie sky. We live halfway in the sky here; in the storms that we watch moving over the land for fifty miles; in the warm blue sky that stretches its vastness from horizon to horizon. This is my home.

"Getting to know a place is a mysterious process. I am of the second generation here. It is hard to know one's place in the land when one has such a drastic effect on its appearance. Yet, though this country is farmland now, there is much here that is unchanged. I think we live in a wilderness. I think that the wilderness can exist in men's minds as much as it exists anywhere. To find our place in this world is to feel our own wildness, to be aware, to catch the songs as they emerge from the spirit, to walk softly through the grasses or on the canyon rim, and feel the exultant wind binding the earth to the sky, to find freedom in a wild land. Our wildness is within us, yet, it is easy to miss or ignore it.

"As a culture, we find ourselves, as if by surprise, in a new land, despoilers of whole peoples and places. The trauma is immense and we strike out, in fear, at the wilderness that surrounds us—reminding us. We cannot see that we ourselves are wild and our violence rebounds against us.

"I occasionally see a tree in the mellow glow of sunset, a tree that I pass unseeing a hundred times a day, and for an instant it is transformed. Each branch and each leaf glows with indescribable color. I see a part of the awesome, unfathomable world that is our home. It becomes clear what a sense of place is: union with all the trees and spirits and the sunset to make the vast openness of our mind and soul equal to the open sunlit sweep of the grassy prairie.

"The tree fades into gathering darkness until just the tips of the branches quiver with a magic luminescence and fade into night.

"In the darkness we stand and listen to the wind."

In Beauty it will be finished.

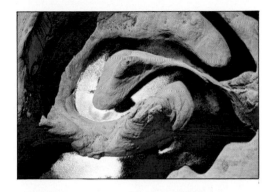

Lake Powell and the flooded end of Escalante.
John J. Reyna, Taos Pueblo.
Drag line at Peabody strip mine on Black Mesa,
Arizona.

Petroglyph with birds in different flight positions, somewhere west of San Rafael Swell, Utah.
Keet Seel ruin at Navajo National Monument, Arizona.
Opposite: Jeannie Moreno in the throne room of "The Grotto," canyonlands.

Ron Warnick, native son.
Student from Evergreen College, Olympia, Washington.
Gary Smith, by Jim Conklin.

SONGS

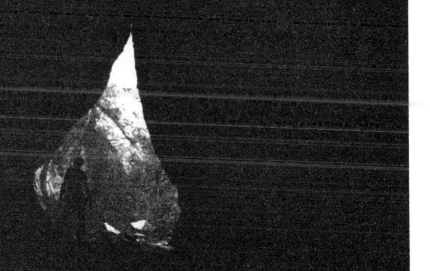

Color Crayon Morning

Words and music by Gary M. Smith

Something Big

Words and music by Gary M. Smith

Cracked side-walks and red fire hy - drants, Street-car trol-ley rol-ling
through the square— Pig - eons a - fly - ing news boy a - cry-ing ——
Some-thing big is mov-ing —— mov-ing through the air! ——

Fine

Morn - ing comes —— to the ci - ty in a flur - ry,
Start now to seek ———— the si - lence of your spac - es,
Can you see that I am run-ning on a rain - bow?

Peo-ple be - gin a - gain to be - gin. ————   Chas - ing
Deep with- in your - self a sym-pho - ny. ————   Fa - ces
Would you like to rise and go with me. ————   Throw a-way your

dreams, plan - ning schemes, For this could be that
pass-ing, In-side your ask-ing, What is this that's
fear, The time is draw-ing near. Won't you try to

spe - cial day, __ A day to look and start to see, — And
Cal - ling me, __ Down boul - e - vards and through the streets,
stop and see, __ That I am you and you are me? __ Through

move up to - ward that some-thing, — Some - thing mov - ing
Rid- ing el - e - va - tors up,
love to geth er we can touch

1. C

big! _____ Climb, Climb, Climb in - to the air! _____

2. C   C   D

3. C   C   D

_____ Touch and catch that some-thing mov - ing big! _____

D.C. al fine

Windsinger

Wind‑sing‑er, ride! Wind‑sing‑er, ride! Wind‑sing‑er ride ____

____ NIT C'‑HI HA‑TA‑TIH, NIT C' HI HA‑TA‑TIH

ride. ____

[sung]	A young Na‑va‑jo came ri‑ding, ____	While at his back set the
[spoken]	He was carried to the North from the mountain,	To the land of the bear and
[spoken]	He was carried by the wind to the Southland,	Where the Lion and the Llama

bleed‑ing, des‑ert sun; ____	He sought his name on the	moun‑tain,
silent, frozen faces; ____	The bear sat and	watched the Eagle,
crawled on bloated bellies; ____	Their eyes were only blackened	sockets,

From the wise man called Na‑ga‑Khan ____	Na‑ga‑Khan was such an
Whose talons had been tangled in the darkness.	The Eagle tried to
And Children tried to run while rickets swelled their knees	The rich would feast and

old man And his eyes were filled with si - lence
stretch his wings And to tear himself free from the darkness;
make their speeches; to morrow would bring the needed changes;

The si-lence passed from the a-ges, — When the spir-its first walked and
But his people were all sleeping, And not one person cared to
But their speeches were all empty, And the hungry children groaned till

breathed up-on the land ———— The old man smiled at the
wake and set him free. The boy breathed deeply from the
their parents could not sleep. Then the groaning changed into a

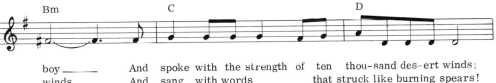

boy ——— And spoke with the strength of ten thou-sand des-ert winds;
winds, And sang with words that struck like burning spears!
roar, The roar began to thunder and the wind began to scream:

"Ride the four winds of the moun-tain; — Wake, and See, and

carried to the East across the ocean,   To a place where the Dragon spewed its
rode among his   people,   Singing in the cities the
as   he rode he heard a   rumb-ling,— For a few had woke   and

venom on the land;   He saw the vacant stares of the
suburbs and the ghettoes;   He   sang to his brothers in the
turned to raise their fa-ces; —— They came forth from the dark-ened

millions,   Hungry stares that held no hope for food in empty hands.
hogans,   In   the squalor   of   the res-er-va-tions.
ghet-toes —   From the   ci-ties, and off the   res-er-va-tions. ——

The Dragon   crouched before the   Eagle,   That   had
But the people   would not   listen;   And   the
A   few had stopped to hear the   sing-ing, ——   And   a -

gorged   itself   on darkness, That had slept   itself   to
children   danced   their frenzied dances, As the neon whirled, and flashed
wak-ened from the slum-ber of the man-y; — Each one raised a burn-ing

weakness,   In whose eyes grew dim the light that could   still make all men free.
and blinded;   And   poison bearing clouds hung like   incense in the air.
spear, ____   With which to pierce the dark-ness and let   forth the   shin-ing light._

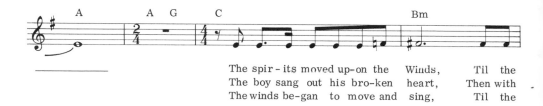

_____   The spir-its moved up-on the   Winds,   Til   the
The boy sang out his bro-ken   heart,   Then with
The winds be-gan to move and sing,   Til   the

voices of the ages began to   echo in his ears: [sung] "Ride a-gain a-mong your
sorrow in his eyes he rode slowly   to the West.
val-leys   e-choed with a   hun-dred rol-ling songs;

peo-ple; ___   Make   them wake   and   see,   Wind-sing-er!"_____

Wind-sing-er ride   Wind-sing-er ride Wind-sing-er   ride. ___

[spoken] Wind-sing-er [sung] "Dance while you sleep my peo-ple;— The

songs that I sing have no mean-ing— for your ears." ———

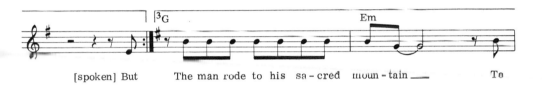

[spoken] But The man rode to his sa-cred moun-tain— To

Wake, and See, and Think, and Speak Wind-sing-er!———

Wind-sing-er, ride! Wind-sing-er ride! Wind-sing-er ride!———

NIT C' HI HA-TA-TIH NIT C' HI HA-TA-TIH ride!———

Dick's Song

Words and music by Gary M. Smith

Carve your rest-less song in me.
Swal-low up our ques-tions, hopes and fears.
cross a land so vast and big, it wan-ders out of
makes us look to see our depths Where - in the Spir - it

I don't know why, Time seems to pass me
the river's pray-in' OM, MANI PAD-ME
sight. Yet the sun still hangs a-round, Til the night-birds cry their
dwells. Like the An-a-sa-zi et-ching, His ghost-ly In-dian

by, Like the riv-er that's rol-lin' Through the
HUM. I won-der where it's rol-lin' me And
sounds Then rest-ing on a Moun-tain-top It
paint-ings He tried to hold death's muf-fled feet And

si - lent shades of life. [yodeled] Yo-del - oo
when the light will come.
lin - gers, then is gone.
Keep his life from fading!

[repeat first verse]

———— , oo ——————————

The Red Tail

Words and music by Gary M. Smith

Chorus

I am the Red Tail,— Climb-in' and soar - in' thru our sky, With the

sun - set__ on my fea - thers,_ With your friends all get-tin' high! _____

High-er ____ than the Red Tail, ____ Up a bove me there is
Like _____ the Red Tail,____ Catch - ing cur - rents, and __

Beau - ty all a - round. Out be - yond me, there breathes Beau-ty, Down be -
ri - sing in the sky. Out a - way from all that's u - gly, Brea-thing

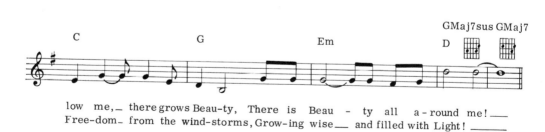

low me,_ there grows Beau-ty, There is Beau - ty all a - round me! __
Free-dom_ from the wind-storms, Grow-ing wise__ and filled with Light! _____

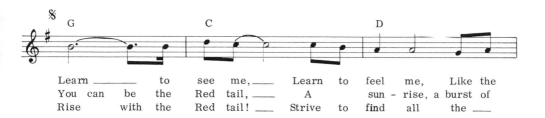

Learn ____ to see me, ____ Learn to feel me, Like the
You can be the Red tail, ____ A sun - rise, a burst of
Rise with the Red tail! ____ Strive to find all the ____

wind a - cross my wings, Let my Spi - rit ____ grow with - in you, ____ Learn to
glo - ry in the sky, You'll know Beau - ty, ____ You'll know Free - dom, ____ You'll find
Beau - ty in your life. Like the Red Tail you caught at sun - set, ____ Like the

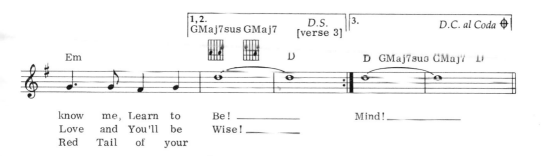

1, 2.
GMaj7sus GMaj7   D.S.   3.   D.C. al Coda
                            [verse 3]   D GMaj7sus GMaj7 D

Em   D   D GMaj7sus GMaj7 D

know me, Learn to Be! ____   Mind! ____
Love and You'll be Wise! ____
Red Tail of your

Coda
GMaj7sus GMaj7

D   C   G

high! ____ all get - tin' high, ____ all get - tin' high! ____